RATE-DISTORTION BASED VIDEO COMPRESSION

Optimal Video Frame Compression and Object Boundary Encoding

RATE-DISTORTION BASED VIDEO COMPRESSION

Optimal Video Frame Compression and Object Boundary Encoding

Guido M. SCHUSTER
U.S. Robotics
Skokie, Illinois, USA

and

Aggelos K. KATSAGGELOS
Northwestern University
Evanston, Illinois, USA

KLUWER ACADEMIC PUBLISHERS
BOSTON / DORDRECHT / LONDON

A C.I.P. Catalogue record for this book is available from the Library of Congress

ISBN 0-7923-9850-5

Published by Kluwer Academic Publishers,
P.O. Box 17, 3300 AA Dordrecht, The Netherlands.

Kluwer Academic Publishers incorporates
the publishing programmes of
D. Reidel, Martinus Nijhoff, Dr W. Junk and MTP Press.

Sold and distributed in the U.S.A. and Canada
by Kluwer Academic Publishers,
101 Philip Drive, Norwell, MA 02061, U.S.A.

In all other countries, sold and distributed
by Kluwer Academic Publishers Group,
P.O. Box 322, 3300 AH Dordrecht, The Netherlands.

Printed on acid-free paper

Printed in the Netherlands

To Dawn

CONTENTS

LIST OF FIGURES

Chapter 5

Chapter 7

Chapter 8

LIST OF TABLES

Chapter 8

PREFACE

One of the most intriguing problems in video processing is the removal of the redundancy or the compression of a video signal. There are a large number of applications which depend on video compression. Data compression represents the enabling technology behind the multimedia and digital television revolution.

In motion compensated lossy video compression the original video sequence is first split into three new sources of information, segmentation, motion and residual error. These three information sources are then quantized, leading to a reduced rate for their representation but also to a distorted reconstructed video sequence. After the decomposition of the original source into segmentation, motion and residual error information is decided, the key remaining problem is the allocation of the available bits into these three sources of information. In this monograph a theory is developed which provides a solution to this fundamental bit allocation problem. It can be applied to all quad-tree-based motion compensated video coders which use a first order differential pulse code modulation (DPCM) scheme for the encoding of the displacement vector field (DVF) and a block-based transform scheme for the encoding of the displaced frame difference (DFD). An optimal motion estimator which results in the smallest DFD energy for a given bit rate for the encoding of the DVF is also a result of this theory. Such a motion estimator is used to formulate a motion compensated interpolation scheme which incorporates a global smoothness constraint for the DVF.

Several algorithms of general nature pertaining to the problem mentioned in the previous paragraph are also presented in this monograph. Among them is an optimal bit allocation scheme for dependent quantizers with arbitrary dependencies, a very fast convex search based on a Bezier curve, an optimal bit allocation scheme for dependent quantization using the minimum-maximum distortion criterion, an optimal scanning path for quad-trees, and an optimal quad-tree decomposition with leaf dependencies.

The optimal bit allocation problem is a rather old one in the information theory community. The optimal bit allocation problem as part of the design of a coder and decoder (codec), or the design of a codec in the rate-distortion (R-D) sense has became a quite popular problem in the last few years. Such a problem becomes extremely "critical" when codecs are designed to operate at low bit-rates, as represented by one of the functionalities of the ongoing standardizations effort MPEG4. Various other related encoding problems in the R-D sense have also become quite popular.

One such problem is the encoding of a boundary by either minimizing the bit rate given an acceptable level of distortion, or minimizing the distortion for a given bit budget. Such a problem plays an important role in object oriented video coding. In this work a polygonal approximation of the original boundary is used and two solutions are developed. With one of them the vertices of the polygon coincide with boundary points, while with the other one they are allowed to be located inside a band formed around the original boundary. Several different classes of distortion measures are investigated which result in different algorithms. Most of these algorithms are based on a weighted directed acyclic graph formulation of the problem.

This monograph was originally written by Guido M. Schuster as a doctoral thesis under the supervision of Prof. Aggelos K. Katsaggelos at Northwestern University. We gratefully acknowledge the financial support from Motorola Inc. and Northwestern University. The discussion with the members of the Motorola Visual Communications Group, especially Steve N. Levine, James C. Brailean, Mark R. Banham, Chueng Auyeung and Kevin O'Connell, have resulted in many important inputs to this research effort. Finally, Guido Schuster wants to thank his wonderful wife, Prof. Dawn Barnes-Schuster. She did not only support him morally but she also spend many nighttime hours with him in the Lab. completing this work.

Guido M. Schuster and *Aggelos K. Katsaggelos*

1

INTRODUCTION

In this chapter we provide some motivation of why video compression is an important problem. We also take a conceptual look at how motion compensated video compression schemes work and we point out some of the shortcomings of traditional approaches. We then introduce the basic idea behind the proposed video compression scheme and give an overview of the monograph.

This chapter is organized as follows: In section 1.1 we provide some motivation of why video compression is an important problem. In section 1.2 we state the general video compression problem for a motion compensated video coder. In section 1.3 we point out the specific and general contributions of this monograph. In section 1.4 we give a overview of the monograph and in section 1.5 we summarize this chapter.

1.1 MOTIVATION FOR VIDEO COMPRESSION

In this section we provide some motivation of why video compression is an important problem. Note that we use the term "compression" as a synonym for source coding, and hence video compression and video coding are equivalent terms in this monograph. Just a few years ago, digital video [1] was only used in professional studios. This was so because large amounts of storage and processing power were needed and the computers required to manipulate video where prohibitively expensive. The personal computer industry has seen an incredible increase in available processing power which has enabled the introduction of desktop video [2], video based interactive multimedia services [3] and desktop video conferencing [139]. With the signing of the new telecommunication bill in the United States of America, the legal groundwork has been laid for the merging of telephones, personal computers and television sets [9]. This merging of the markets will result in fierce competition between telephone companies, cable TV providers, TV manufacturers and computer manufacturers as predicted in [201].

One of the basic ingredients of this brave new multimedia world are digital images and digital video, since a picture is worth a thousand words. In fact, the recent boom in the Internet was mostly spawned by the introduction of the World Wide Web (WWW), which enabled for the first time the reception of digital images and videos on demand, without special equipment or knowledge requirements by the user.

Besides the emerging multimedia market, the change from analog to digital video for broadcasting [26, 155, 108] is only possible because of video compression technology. The Federal Communications Commission (FCC) requires that the High Definition TV (HDTV) transmitted signal fits inside the same 6 Mhz channel spacing as is currently used for today's NTSC transmissions. This allows the FCC to more easily assign available channels to the HDTV service. This requirement implies that 1.5 Gbits/second (HDTV) of video data need to be compressed and modulated such that they fit into the available 6 Mhz channel which can support about 20 Mbits/second.

One of the enabling technologies behind the multimedia and digital TV revolution is data compression. This is because the physical channels employed have usually not been designed for the bandwidth required by multimedia applications. For example, one of the most desired products is a video phone which uses the public switched telephone network (PSTN, sometimes also called POTS for "plain old telephone service") [139, 104, 78]. These networks have originally been designed to transmit analog speech signals and not digital video. The fastest modems available today for PSTN can transmit at a bit rate of 28.8 kbits/second [94] (for very good connections, 33.6 kbits/second modems have just arrived on the market). Many consider this the upper limit for voiceband modems over PSTN since the information theoretical capacity $C = W * \log_2(1 + \text{SNR})$ [57, 70] is around 30 kbits/second for a standard telephone line (Bandwidth: W = 3 kHz, Signal to noise ratio: SNR = 30 dB). Hence if we would like to use a video phone over these existing lines, then the maximum bit rate available is 28.8 kbits/second.

A common format for such an application is the quarter common intermediate format (QCIF) which is of dimensions 176×144 pixels. Consider an original

QCIF color sequence which is stored with 24 bits per pixel (bpp), 8 bpp for the red, green and blue channel. If we further assume that the frame rate is 30 frames/second then this results in a required bit rate of 176*144*24*30 = 18.3 Mbits/second. This shows that the uncompressed (original) sequence requires 634 times larger bandwidth than what is available using PSTN. This example clearly illustrates the importance of video compression. As we will see later on, the video coder we develop in this monograph, is capable of transmitting QCIF video at bit rates below 28.8 kbits/second.

Even though video over telephone lines is an extreme example of a mismatch between the required bandwidth of the video source and the available bandwidth of the channel, there is almost never enough bandwidth available to send an uncompressed video sequence. Alternatively, instead of sending the original video sequence at a given spatial and temporal resolution with no compression at all, one can send a higher quality video sequence, i.e., one with higher temporal and/or spatial resolution, over the same channel using video compression techniques. Clearly the customer will appreciate the better picture and the smoother motion. The drawback of video compression is the computing power required to encode and decode the sequence but the advantage is that for a given channel capacity, the best possible video sequence is transmitted.

1.2 PROBLEM STATEMENT

In this section we state the general video compression problem for a motion compensated video coder. We point out the shortcomings of the traditional approaches and introduce the basic idea behind the proposed approach.

A straightforward video compression scheme is to apply an image compression scheme, such as JPEG [216, 154], to each frame of the video sequence. In fact this is a well know scheme sometimes referred to as motion JPEG or M-JPEG. Image compression is based on the fact that every natural image exhibits a high spatial correlation among neighboring pixels and that a it contains a significant amount of perceptually irrelevant information. In video compression, an additional correlation along the temporal axis can be exploited, i.e., both, the spatial and temporal neighbors contain information about the current pixel.

The main idea behind motion compensated video compression is to use the temporal and spatial correlation such that the current frame can be predicted as accurately as possible from the previously decoded frames. The more accurate the prediction is, the smaller is the entropy of the prediction error, which is usually transmitted using a lossy compression scheme, which removes the perceptual irrelevant information. If the entropy of the prediction error is small, then only a small amount of information needs to be transmitted and hence the compression ratio is high.

Most video compression schemes utilizing the temporal correlation use the concept of motion compensation. The frame is segmented into objects, where the pixels inside the objects are assumed to move according to a motion model employed for this object and the parameters which specify the model. The motion and the segmentation is then used to predict the current frame from the previously reconstructed frame. Since this prediction fails in some regions, such as, uncovered background and areas where new objects are introduced, the residual error has to be transmitted. The residual error is the difference between the current frame and the predicted frame. In Fig. 1.1 this motion compensated video coding scheme is displayed.

Fig. 1.2 shows the decomposition of the original sequence into three new pieces of information, the segmentation, the motion and the residual error. The main motivation for splitting the original sequence into these three parts is the assumption that the sum of the individual bit rates, for representing each piece of information, is smaller than the bit rate required to represent the original sequence. Natural video sequences exhibit high temporal correlations and therefore this assumption holds true in general.

So far the video compression scheme is lossless which means that the reconstructed sequence is identical to the original sequence. In a lossless scheme, the compressed video sequence contains exactly the same amount of information as the original sequence. In a lossy scheme information loss occurs and the compressed video sequence is not identical to the original sequence. The distinction between these two approaches is of fundamental importance, since conceptually for a lossy scheme any compression ratio can be achieved, by losing more and more information, whereas in a lossless approach the minimum bit rate for an information source is given by its entropy [57]. Note that the entropy of a given source is unknown but can be bounded by assuming certain statistical models.

Lossy schemes are commonly used for the encoding of video sequences since they achieve a much higher compression ratio than lossless schemes. The reason for this is twofold. First, the sensors used to acquire a given sequence might result in a spatial and/or temporal resolution which is better than what the human visual system (HVS) can perceive. Therefore information which is contained in the sequence but is not perceived by a human observer can be

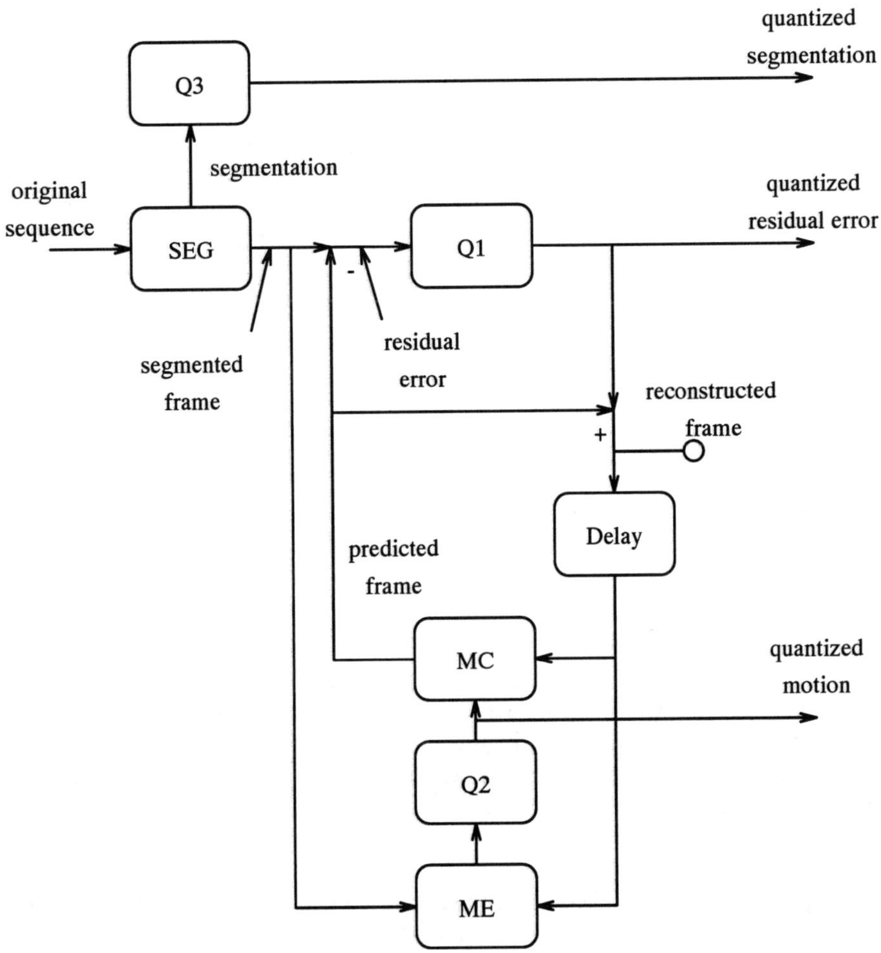

Figure 1.1 Block diagram of a motion compensated video coder. Q1: quantizer for the residual error; ME: motion estimation algorithm; Q2: quantizer for the motion; SEG: segmentation algorithm; Q3: quantizer for the segmentation; MC: motion compensation algorithm; Delay: frame delay.

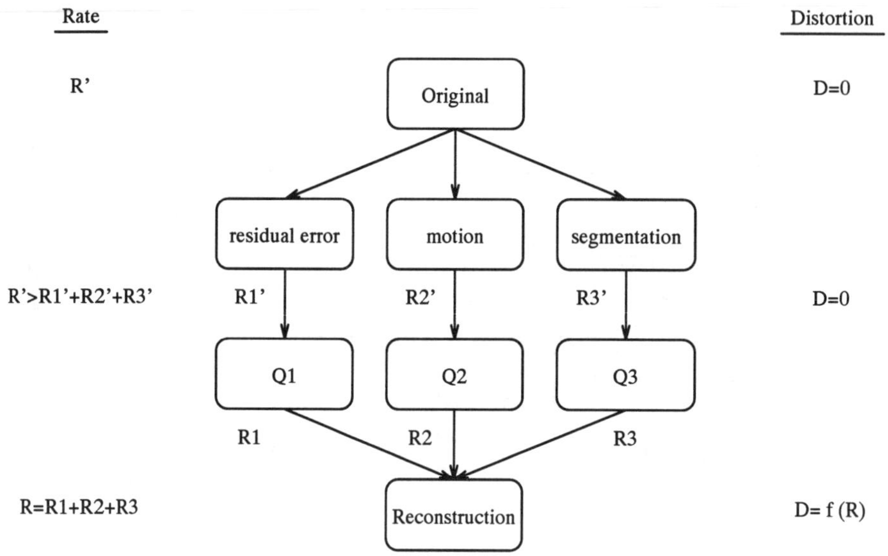

Figure 1.2 Decomposition of the original sequence into segmentation, motion and residual error.

eliminated. A good example of this is the commonly used sub-sampling of the color components since the HVS spatial resolution for color is much smaller than that for luminance. Another example is the finer quantization of the low frequency components than the high frequency components in JPEG. Second, the bit rate required for encoding a certain element of the sequence might be quite high, though the perceived distortion caused by not sending this element is rather small. An example would be a very "busy" texture, which requires a high bit rate, although an error in this texture is almost invisible. Clearly the high compression ratio of a lossy scheme comes with a price, namely that the compressed video sequence is not identical to the original one. The difference between the original sequence and the compressed sequence is usually measured by a distortion measure such as the mean squared error. For every lossy video coding scheme, there exists an inherent trade-off between the rate (compression ratio) and the resulting distortion [13]. A low distortion requires a high rate and vice versa a low rate results in a high distortion. We will formalize these concepts mathematically later on, when we introduce the elements of operational rate distortion theory used in this monograph.

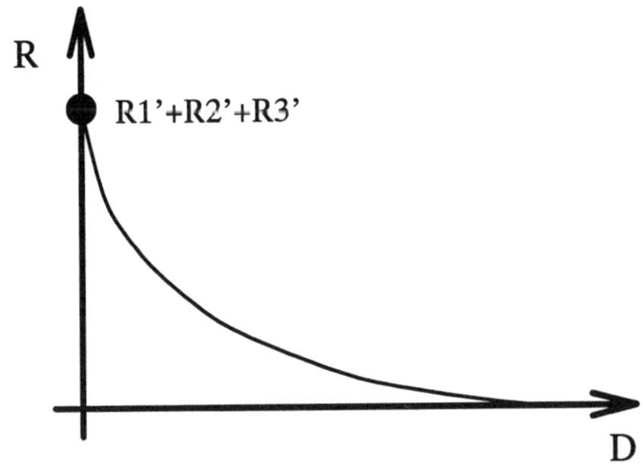

Figure 1.3 Operational rate distortion curve of a motion compensated video coding scheme.

Lossy video compression schemes are more common since they can achieve a much higher compression ratio than lossless schemes. The information loss occurs during the quantization of segmentation, motion and residual error, and the difference between the original and the reconstructed video sequence can be accessed by a distortion measure $D = f(R)$, where D represents the distortion, R the required rate and $f(\cdot)$ the distortion rate function, which is the inverse of the rate distortion function introduced later on.

A fundamental problem of a motion compensated video compression scheme, which is not addressed by the traditional approaches, is the bit allocation among the segmentation, motion and residual error. This can be stated as follows,

$$\min D, \qquad \text{subject to:} \qquad R \leq R_{max}, \tag{1.1}$$

where the distortion D and the rate R depend directly on the bit allocation among segmentation, motion and residual error. The optimal solution to the above constrained optimization problem, for any given maximum rate R_{max}, results in the operational rate distortion curve of a motion compensated video coding scheme, which is displayed in Fig. 1.3. Every point on the operational rate distortion curve in Fig. 1.3 represents the optimal bit allocation among segmentation, motion and residual error. In other words, for a given rate R_{max} there exist many different segmentations, motion parameters and residual error encodings, but only the one which results in the smallest distortion is picked.

1.2.1 Traditional approach

Many motion compensated video compression schemes have been proposed in the past, which all take advantage of the spatio-temporal correlation and the perceptual irrelevant information inherent in a video sequence. In chapter 2 we present a review of the most popular lossy video compression schemes.

Video compression can be split into three parts. The first part is the definition of the problem, the second part is the development of a mathematical model and the third part is the optimization of the parameters used in the model. The definition of the video compression problem has to take into account the rate distortion performance, the complexity of the encoder and the coding delay. It should also contain the measure which will be used to evaluate the final compression scheme and the class of sequences for which the system is designed. The second part, the development of the mathematical model, which are the methods used to efficiently encode a video sequence, has been extensively studied in the literature. In traditional approaches, these methods, such as vector quantization or motion compensated video coding, are usually the focus of the scheme. Note that if there is not a clear definition of the video coding problem, i.e., part one was not given enough attention, than it is very hard to evaluate the performance of these methods. The third part, the optimization of the employed methods, is usually referred to as rate control or bit allocation. In traditional approaches, little attention is paid to the optimal allocation of bits, but heuristic rules are employed. Note that in a scheme where the methods are not used in an optimal fashion, the resulting overall video coder will not perform at its best. Furthermore, the video coding problem needs to be clearly defined, otherwise it is not clear what exactly is the objective of the optimization.

The rate distortion framework is well established to measure the performance for a certain scheme, when the coding delay is fixed. In addition, many efficient methods for video compression have been developed. In contrast to the first two parts, which are well developed, the optimal application of the methods, i.e., the optimal bit allocation problem, has not received the attention it deserves.

As mentioned in the previous section, the most prominent video coding approach is motion compensated video coding which splits the original sequence into three pieces of information, segmentation, motion and residual error. In traditional schemes, very little attention is paid to the allocation of the available bits among these sources of information, even though this is a fundamental problem of motion compensated video coding.

There are three main tasks which have to be accomplished, the segmentation of the scene, the estimation of the motion and the quantization of the residual error. In traditional approaches, three different and independent methods are used for these tasks. The only task which even considers the resulting rate distortion tradeoff (which is the performance measure) is the selection of the quantizers used for the quantization of the residual error. When rate control is employed, the quantizers are selected so that a target rate can be achieved. Clearly the resulting bit allocation among motion, segmentation and residual error is ad-hoc and by no means optimal. In most cases finding the optimal solution would be much too costly with respect to the time complexity of the algorithm. The main reason for this is that during the modeling of the problem, the designer did not take into account the complexity of the required optimization approach.

1.2.2 Proposed approach

We propose a video coding scheme where the bit allocation among motion, segmentation and residual error is accomplished in an optimal fashion. In this monograph we examine all three parts of the video coding problem together.

For the first part, the definition of the problem, we employ the established rate distortion framework. In other words, we examine the resulting distortion for a given bit rate to judge the performance of the scheme, or vice versa, we examine the resulting bit rate for a given distortion. Other performance measures we employ are concerned with the speed of the resulting video coder and are based on complexity theory. We will develop efficient algorithms, i.e., the computational complexity is low, but the primary performance measure is the rate distortion tradeoff.

Besides clearly defining the employed performance measure, we differ from the traditional approaches in that the mathematical model we create is strongly linked to the established performance measure and to the necessary optimization procedure. In other words, we develop the video coder such that the optimal decision strategy can be found employing a low complexity optimization approach. Note that there is an art in creating a model so that it is powerful enough to achieve high performance, but also simple enough to result in efficient optimization procedures.

The proposed video compression scheme is a quad-tree based motion compensated video coder where the bit allocation between segmentation, motion and residual error is optimal. We demonstrate the effectiveness of the proposed video compression scheme by comparing it to the current very low bit rate video coding standard H.263, and we show that it can outperform H.263 in the rate distortion sense. We also show the efficiency of the optimization algorithm by deriving the computational complexity and comparing it to an exhaustive search.

1.2.3 Scope of the monograph

Digital video compression contains many different fields, from Digital Signal Processing to Information Theory. In this monograph we are mostly concerned with the optimal bit allocation among segmentation, motion and residual error for a motion compensated video coder. The optimality is in the operational rate distortion sense, i.e., we attempt to find for a given video compression scheme and a given bit rate the decisions which will result in the smallest distortion. We also search for new video compression schemes which lend them self to efficient optimization, but are still powerful enough to result in excellent compression. The bit allocation is carried out optimally for each frame of the sequence, and we do not attempt to optimally allocate the bits among the frames of a sequence. Therefore no buffer control aspects of video coding are discussed in this monograph, though the presented video scheme is well suited for any of the existing buffer control strategies. Throughout the theoretical parts we assume that a finite set of admissible quantizers is given and that the entropy codes are known. The quantizer design and the entropy code design is not a focus of this monograph and in the experimental results we present, the quantizers and entropy codes are taken from the literature.

1.3 CONTRIBUTIONS

In this section we point out the general and specific contributions of this monograph which are mostly in video and segmentation compression and in operational rate distortion theory. Some of these contributions are of very general nature and can be applied to other problems than the ones discussed in this monograph. The general contributions which are discussed in chapter 4, are:

- The derivation of an optimal bit allocation scheme for dependent quantizers with arbitrary dependencies, using the total distortion criterion.

- The development of a very fast convex search for the optimal Lagrangian multiplier λ, which is based on a second order Bezier function.
- The derivation of an optimal bit allocation scheme for dependent quantizers with arbitrary dependencies, using the minimum maximum distortion criterion.
- The development of an optimal scanning path for a quad-tree decomposition which is based on a Hilbert curve.
- The derivation of an optimal quad-tree decomposition with leaf dependencies, for the total minimum distortion criterion and the minimum maximum distortion criterion.

The above contributions have lead to the development of:

- An optimal motion estimation scheme in the rate distortion sense.
- A motion compensated interpolation scheme with a global smoothness constraint for the displacement vector field.
- A video coding scheme with optimal bit allocation between motion and residual error.
- A video coding scheme with optimal bit allocation among segmentation, motion and residual error.
- An optimal segmentation encoding scheme in the rate distortion sense.

which constitute the specific contributions of this monograph.

1.4 OVERVIEW

In this section we give an overview of the monograph. In chapter 2 we review the field of lossy video compression. In chapter 3 we introduce the concept of operational rate distortion theory and the mathematical tools used in this monograph, which are the Lagrangian multiplier method and dynamic programming. In chapter 4 we describe the general contributions of this monograph to the field of operational rate distortion theory. In chapter 5 we propose a motion estimation scheme which is based on one of the general contributions. We show

how the general scheme can be used to build an optimal motion estimator in the rate distortion sense. In that chapter, we also introduce a motion compensated interpolation scheme which is based on the proposed motion estimator. The motion compensated interpolation scheme results in a globally optimal tradeoff between the smoothness of the displacement vector field (DVF) and the energy in the displaced frame difference (DFD). In chapter 6 a fixed block size based video compression scheme is introduced which is based on the proposed motion estimator. The video compression scheme results in an optimal bit allocation between DVF and DFD. In chapter 7 a variable block size based video compression scheme is presented which is an extension of the fixed block size scheme. This scheme results in the optimal bit allocation among motion, segmentation and residual error. In chapter 8 we present a segmentation encoding scheme which is optimal in the rate distortion sense. We develop the scheme for two different classes of distortion measures, where the first class is based on the maximum operator and the second class is based on the summation operator.

1.5 SUMMARY

In this chapter we provided some motivation of why video compression is an important problem. We described a general motion compensated video compression scheme and we pointed out some of the shortcomings of traditional approaches. We then introduced the basic idea behind the proposed video compression scheme and gave an overview of the rest of the monograph.

2

REVIEW OF LOSSY VIDEO COMPRESSION

In this chapter we review the most popular lossy video compression schemes. There have been several excellent reviews published in the past [133, 85, 125, 52, 101] and this review is strongly influenced by these publications. In addition to these publications there exist many different books which contain introductory reviews of this subject [202, 212, 15]. Whenever appropriate, we point out where the contributions of this dissertation fit into the establish body of knowledge.

This chapter is organized as follows: In section 2.1 we point out the difference between lossy and lossless schemes and then focus on lossy video compression since for most practical schemes, the required compression ratios can only be reached by discarding information. In section 2.2 we introduce the paradigm of motion compensated waveform coding and discuss the most popular methods to estimate the motion and to encode the residual error. In section 2.3 we look at three dimensional waveform coding and discuss the three most common approaches. In section 2.4 we present model-based video coding schemes which employ complex scene models to represent the video sequence and in section 2.5 we summarize this chapter.

2.1 LOSSLESS VERSUS LOSSY COMPRESSION

In this section we discuss the difference between lossless and lossy video compression schemes. A lossless video compression scheme trades processing power for capacity, i.e., information is stored or transmitted in a different but equivalent form which is more compact. Since the processing power for the compres-

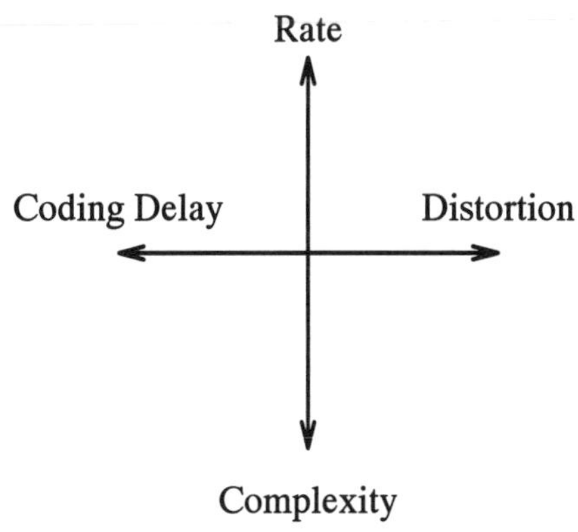

Figure 2.1 Tradeoff space of a video compression scheme.

sion and decompression step is available, less memory or bandwidth is needed for the storage or transmission of a digital movie.

The entropy of a source is the information theoretical limit for lossless compression [188]. Any encoder which wants to compress below the entropy of the source, has to be lossy and therefore some sort of distortion has to occur. Rate distortion theory defines a curve, called the rate distortion function, which is an extension of the concept of entropy when distortion is permissible [13]. The rate distortion function is the lower bound for the rate, given a certain distortion, regardless of the coding scheme employed.

Every video coding scheme involves a tradeoff along the four dimensions shown in Fig. 2.1 [15]. This research is mainly concerned with the rate distortion relationship and the computational complexity of the proposed system. The coding delay is not explicitly addressed, although the proposed solution requires a coding delay of one frame, which is the smallest frame delay possible for a scheme which employs the temporal correlation between frames. A small coding delay is very important for interactive services, such as, video telephony or multimedia database retrieval.

There is an inherent tradeoff between the distortion and the rate of a lossy video coding scheme. A decrease in distortion can only be achieved by an increase in rate, and an increase in distortion will lead to a decrease in rate. We will see that this observation is the basic concept behind the proposed video coding scheme.

2.2 MOTION COMPENSATED WAVEFORM CODING

In this section we discuss motion compensated waveform coders. First we present popular motion estimation schemes for video coding. Then we discuss different schemes for encoding the residual error and finally we present a generic motion compensated transform coder which uses the discrete cosine transform (DCT) for the encoding of the residual error. This generic coder is the basis for all existing standards, such as H.261, H.263, MPEG-1 and MPEG-2.

As we pointed out in the introduction, the main reasons why digital video lends itself to compression are the high spatial and temporal correlations and the perceptual redundancies. Motion compensated (MC) video compression is the paradigm most often used to efficiently employ the temporal correlation. In MC coding, the current frame is predicted from the previous frame using motion information, which has to be estimated first. In this section we assume that the motion information is given as a displacement vector field (DVF). A DVF is a set of motion vectors which determine for each pixel in the current frame their location in the previous frame.

Mathematically, an MC video coder can be described as follows, Let $f_k(\vec{r})$ be the current frame, $\vec{d_k}(\vec{r})$ the DVF and $\hat{f}_{k-1}(\vec{r})$ the previously reconstructed frame which will be used to predict the current frame. Note that $\hat{f}_{k-1}(\vec{r})$ does not need to be a frame from the past, but as in MPEG-2, this could be a future frame when backward prediction is used. The predicted (or motion compensated) frame $\tilde{f}_k(\vec{r})$ is defined as,

$$\tilde{f}_k(\vec{r}) = \hat{f}_{k-1}(\vec{r} - \vec{d_k}(\vec{r})), \tag{2.1}$$

and the displaced frame difference (DFD) $(f_k^{DFD}(\vec{r}))$ is defined by

$$f_k^{DFD}(\vec{r}) = f_k(\vec{r}) - \tilde{f}_k(\vec{r}). \tag{2.2}$$

In a lossy MC coder, the DFD is quantized, that is,

$$f_k^{QDFD}(\vec{r}) = Q[f_k^{DFD}(\vec{r})], \tag{2.3}$$

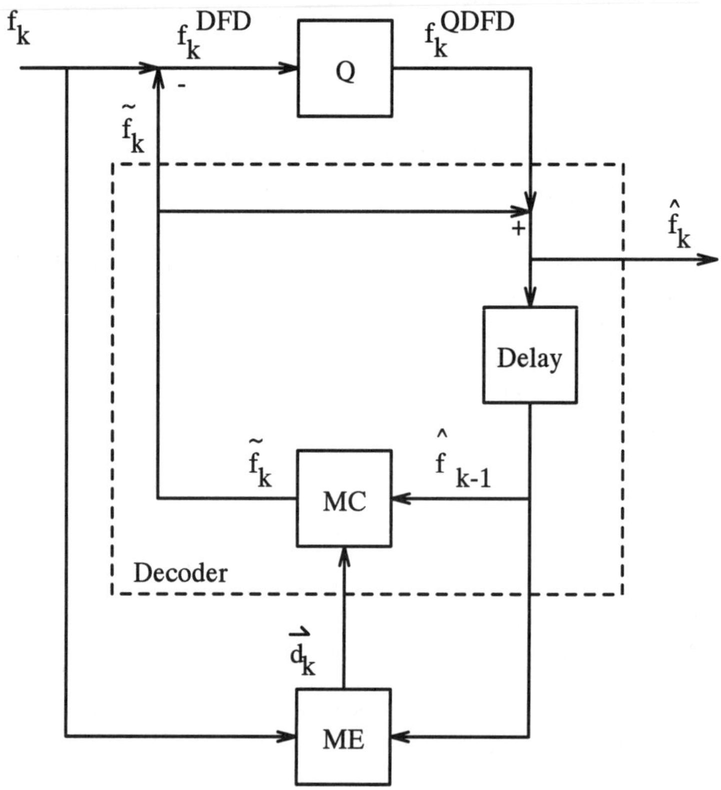

Figure 2.2 A generic motion compensated video coder. Q: quantizer; Delay: frame delay; MC: motion compensation algorithm; ME: motion estimation algorithm.

where $Q[\cdot]$ is the quantization operator. Finally, the reconstructed frame $\hat{f}_k(\vec{r})$, is defined by,

$$\hat{f}_k(\vec{r}) = \tilde{f}_k(\vec{r}) + f_k^{QDFD}(\vec{r}). \tag{2.4}$$

In Fig. 2.2 a generic MC coder and the above introduced notation is displayed. Note that in every MC scheme, the decoder is a part of the encoder, which is indicated in Fig. 2.2. The input to the encoder is the original sequence, whereas the input to the decoder is the quantized DFD and the DVF. In other words, the quantized DFD and the DVF have to be transmitted. Before they are transmitted they are entropy coded using a Huffman code [81] or an arithmetic coder [227].

2.2.1 DVF estimation

The goal of motion estimation in the video compression context is to find a DVF which can be used to predict the current frame as accurately as possible. This is in strong contrast to the general field of motion estimation, where the primary goal is to find a motion field, which is as close as possible to the two dimensional projection of the true three dimensional motion. The latter is a much harder problem than the former. In this section we assume that the video sequence is transmitted using a progressive format. When an interlaced format is used, the motion estimation problem changes slightly [141]. Note that the block labeled "ME" in Fig. 2.2 represents the motion estimation part of the MC coder.

Block matching

Block matching (BM) is by far the most popular motion estimation technique for video coding [87]. The main reason for its popularity is its simplicity and the fact that several VLSI implementations of BM algorithms are available. BM is an optical flow based motion estimator which finds a sparse motion vector field for a block of pixels. Consider Fig. 2.3; the current frame f_k is divided into non-overlapping blocks $B_{\vec{r}}$, and we assume that all the pixels in a block have the same displacement $\vec{d}_k(\vec{r})$. For each block, blocks in a certain search window $W_{\vec{r}}$ in the previously reconstructed frame $\hat{f}_{k-1}$ are compared to the current block and the one which leads to the best match is selected. Mathematically this can be stated as follows,

$$\hat{d}_k(\vec{r}) = \arg \min_{\vec{d}_k(\vec{r}) \in W_{\vec{r}}} \left\{ \sum_{\vec{z} \in B_{\vec{r}}} M(f_k(\vec{r} + \vec{z}), \hat{f}_{k-1}(\vec{r} - \vec{d}_k(\vec{r}) + \vec{z})) \right\}, \tag{2.5}$$

where $\hat{d}_k(\vec{r})$ is the estimate of the DVF for all pixels of block $B_{\vec{r}}$, and $M(\cdot,\cdot)$ denotes the matching criterion.

The main difference between BM algorithms lies in the matching criterion $M(\cdot,\cdot)$ and the search scheme employed to find the minimum in Eq. (2.5). The accuracy of the resulting DVF depends on the matching criterion and it has been shown in [63] that the squared error criterion ($M(a,b) = (a-b)^2$) results in the best DVF estimates. For a VLSI implementation, the square operation gets usually replaced by an absolute value operation, which can be accomplished much more efficiently, since it does not require a multiplication but only a sign check. When a full search (within the search window) is computationally too expensive, faster, but suboptimal approaches can be used [103, 202].

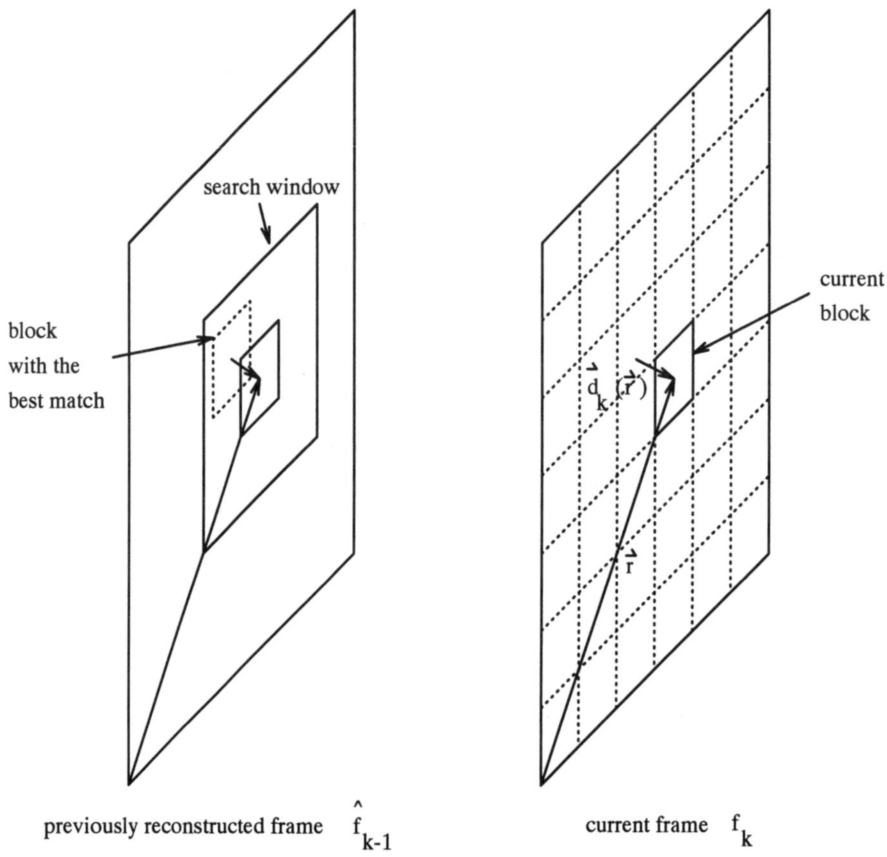

Figure 2.3 Block matching for estimating the DVF between the previously reconstructed frame and the current frame.

After BM, each block in the current frame has a motion vector assigned to it, which points to a block in the previous frame representing the best match. The set of these vectors is called the sparse motion vector field and this field has to be transmitted to the decoder. One of the main problems of BM is the assumption that all pixels in a block move in the same direction. This implies that BM will only do a good job for translational motion and it will fail completely when the block covers two objects which are moving in different directions. Nevertheless, some version of BM is used in almost every implementation of existing video coding standards, such as, MPEG-1 [82, 98], MPEG-2 [83, 99], H.261 [84, 107] and H.263 [46].

Besides the matching criterion and the search scheme, the size of the blocks has a major impact on the quality of the resulting DVF. It is well known [16] that the block size controls the tradeoff between minimizing the matching criterion and finding a consistent DVF. A large block size results in consistent DVF estimates, while a small block size results in smaller matching errors. In order to have the advantages of both, a hierarchical BM scheme [16] can be employed. In such a scheme, blocks of decreasing size are used in succession to find the best motion vector.

Variable block size BM Current standards are all based on a fixed segmentation of the frame into blocks of a given size. The main advantages of this are simplicity and robustness of the algorithm and the fact that no segmentation information needs to be transmitted. The only exception to this is H.263 when the "Advanced Prediction Mode" is used. Then there is the option to split the 16×16 block into four 8×8 blocks. The main disadvantage of a fixed and arbitrary segmentation is that it can not represent the real objects in the scene, and hence the representation of the scene is not as compact as it could be. In addition, the underlying motion model of simple translation cannot capture more complex motion, such as, pan, rotation and zooming. On the other hand, a dense DVF implies also a fixed segmentation into blocks, but there is only one pixel per block. Clearly a dense DVF can represent more complicated motion. The problem, however, in this case is that in many regions of the scene, most of the motion vectors are identical and therefore not needed to describe the motion accurately.

Hence a natural DVF is inherently inhomogeneous, in the sense that for regions with simple motion, one motion vector is sufficient, whereas in regions with complex motion the DVF should be denser. If the scene is segmented into blocks of different sizes, then a compact representation of the DVF can be achieved.

Since the segmentation information of the frame needs to be transmitted, an efficient method should be used to encode it.

The quad-tree (QT) data structure [174] is commonly used to decompose a given frame into blocks of different sizes, since it enables an efficient representation of the resulting decomposition. A QT starts with a square block with side length a power of two. This block can be split into four equally sized square sub-blocks, and only one bit is required to encode the split or lack of it. Clearly this splitting can then be recursively applied to the sub-blocks until a sub-block is of dimensions 1×1. Hence the entire segmentation of the frame can be represented by a tree structure, where every node has four children (hence this structure is called a quad-tree).

The QT structure is an efficient way of segmenting the frame into blocks of different sizes. It is therefore very attractive for the efficient representation of an inhomogeneous DVF. In [136, 11, 96], the QT structure is used to compactly represent the DVF. In [136] the dense motion field is represented efficiently using higher order motion models, and the spatial density of the applicability of these models is encoded using a QT. In [11] the temporal gradient of a pel-recursive motion estimation algorithm is QT encoded and in [96] a QT-based motion estimator is proposed which finds the best QT in the rate distortion sense. Note that the algorithm proposed in [96] is a direct application of the more general scheme proposed in [198, 200]. The motion estimator we propose in chapter 5, is more general than the scheme in [96], since it finds the optimal QT decomposition and the optimal DVF, given that the DVF is first order differential pulse code modulation (DPCM) encoded. The algorithm presented in section 4.5 is a generalization of the scheme presented in [198, 200], since the optimal QT decomposition can be found even when there are dependencies between the QT leafs.

Note that the variable block size based decomposition of the frame is a good compromise between the overly complex object-oriented approaches, which will be discussed in section 2.4.1, and the overly simple fixed block size based schemes. The object-oriented approaches suffer from the fact that it is in general very hard to accomplish joint segmentation and motion estimation while the fixed block size schemes suffer from the inability to represent complex motion. In addition, the overhead paid for the segmentation of a QT-based scheme is much lower than the bit rate required to transmit the fine segmentation of an object-oriented scheme, although it is still substantial considering the fact that a fixed segmentation based scheme does not require any segmentation information to be transmitted.

In chapter 5 we propose a new variable block size based motion estimator which results in an optimal tradeoff between the bit rate required for the DVF and the resulting matching error for the DFD. The proposed scheme is based on variable block sizes and therefore the resulting DVF is inhomogeneous. Depending on the available bit rate, the denseness of the DVF changes, such that for a high bit rate the DVF is quite dense, whereas for a low bit rate the DVF is quite sparse. Since the inhomogeneous DVF represents the real motion quite well, we also propose in chapter 5 a motion compensated interpolation scheme which is based on this motion estimator. The resulting scheme has the advantage that it offers globally optimal control of the tradeoff between a global smoothness constraint for the DVF and the resulting interpolation error energy.

In chapter 7 we present a QT-based video coding scheme which optimally distributes the available bits among the three pieces of information which need to be sent, i.e., segmentation, motion and residual error. The scheme is based on a very efficient representation of the motion using the QT-based motion estimator. It finds the motion and the segmentation jointly as the globally optimal solution of a discrete constrained optimization problem.

Region matching

A generalization of block matching can be achieved by considering more complex motion models than simple translational displacement and different shapes than just rectangular blocks. Common motion models are the affine transform, the perspective transform and the bilinear transform [130]. The affine transform has six degrees of freedom and hence a triangle, which has also six degrees of freedom, is the region of choice, since any two triangles can be mapped onto each other using the affine transform. The perspective transform has eight degrees of freedom and therefore the regions under consideration are usually quadrilaterals. The perspective transform describes the motion of a rigid planar patch in the three dimensional real world, when perspective projection onto the image plane is assumed [80, 209]. The affine transform describes the motion of a rigid planar patch in the three dimensional real world, when orthographic projection onto the image plane is assumed [128]. The affine transformation can be considered an approximation to the perspective transform, since for patches which are far enough away form the focal point and the image plane, perspective projection results in almost parallel rays. The bilinear transform has also eight degrees of freedom and hence its regions are usually quadrilaterals also.

There are several different ways of estimating the motion parameters for these transforms. The obvious choice is an exhaustive search around an initial guess in the six or eight dimensional parameter space which is proposed in [186]. The goal of this search is to find the parameters which minimize the energy of the DFD. Affine motion models have received considerable attention since only six parameters need to be estimated. In [24] a gradient based estimation is proposed which finds a local minimum of the DFD energy. This optimization is initialized using motion vectors found by block matching. In [128] a fast hexagonal matching is proposed which also uses an initial estimate of the motion vectors which is found by block matching. In hexagonal matching, the middle point of a hexagon formed by six equilateral triangles is perturbed with the purpose of minimizing the DFD energy over the hexagon. This procedure is performed iteratively in a raster scan until it converges to a local optimum. In [129] a video coder which employs this motion estimation scheme is presented where the resulting DFD is encoded using a discrete cosine transform. A scheme where the bilinear transform is used for motion compensation based on a fixed mesh is presented in [199]. In [193] the six parameter affine model and a four parameter "quasi-affine" model introduced in [135, 95], are used in a region-based motion estimator.

Adaptive region matching For the above discussed schemes, the original frame is segmented into quadrilateral regions (eight parameter model) or triangles (affine motion model). As for block-based approaches, this segmentation is usually fixed and therefore arbitrary. A logical extension of this scheme is based on an adaptive mesh which segments the scene according to its content [137]. In this scheme quadrilateral regions are used which are adapted to the scene using the two previously reconstructed frames. Hence no information about the mesh locations needs to be sent. The motion compensation is then achieved by warping the previous frame according to the motion information of the mesh control points. A more general scheme, called an active mesh has been introduced in [217]. This scheme results in a deformed mesh which tries to track features, such as edges, and very good prediction performance is reported.

Spatial and temporal gradient methods

Another class of motion estimation algorithms are based on the so called spatio-temporal constraint equation [103]. We assume for this derivation that a video sequence is a space-time continuum where the intensity at a particular point in space $(x(t), y(t))$ and time (t) is denoted by $f(x(t), y(t), t)$. The spatio-temporal constraint equation can be derived by making the following two assumptions. First, the intensity of a point is constant along its trajectory, or in other words, the derivative of the intensity with respect to time is zero. Second, the motion of the point is uniform, i.e., it does not accelerate. This is sometimes referred to as linear motion. By applying the chain rule to the derivative of the intensity the following expression results,

$$\frac{df(x(t),y(t),t)}{dt} = \frac{\partial f(x(t),y(t),t)}{\partial x(t)}\frac{\partial x(t)}{\partial t} + \frac{\partial f(x(t),y(t),t)}{\partial y(t)}\frac{\partial y(t)}{\partial t} + \frac{\partial f(x(t),y(t),t)}{\partial t} = 0. \quad (2.6)$$

For non-accelerated motion, the velocity in the x-dimension v_x equals $\frac{\partial x(t)}{\partial t}$ and the velocity in the y-dimension v_y equals $\frac{\partial y(t)}{\partial t}$. The x and y displacements of the DVF can be found by multiplying these velocities by the time difference Δt between two frames under consideration. Hence the spatio-temporal constraint equation is of the following form,

$$\frac{\partial f(x(t),y(t),t)}{\partial x(t)}v_x + \frac{\partial f(x(t),y(t),t)}{\partial y(t)}v_y + \frac{\partial f(x(t),y(t),t)}{\partial t} = 0. \quad (2.7)$$

Eq. (2.7) expresses the fundamental relationship between the spatial gradients, the velocities and the temporal gradient. In motion estimation, the goal is to estimate the displacement at every location, or in other words, we would like to use this equation to estimate v_x and v_y. Clearly it is impossible to solve Eq. (2.7) for v_x and v_y, since we have two unknowns but only one equation. By assuming that v_x and v_y are constant in a small neighborhood, Eq. (2.7) can be used to build an over-determined system of linear equations, which can be solved using the classical least squares approach.

One problem with Eq. (2.7) is that we assumed that the spatial and temporal gradients are available. Since digital video is discrete in time and space, these gradients need to be estimated. There are many different ways of estimating derivatives [59, 158], but most often finite differences are used.

Besides the above mentioned least squares approach, there have been many different schemes proposed which explicitly or implicitly use Eq. (2.7) [4]. In the case of video coding, one important aspect of the motion estimation procedure is the bit rate required to transmit the DVF. The bit rate required to transmit a dense DVF, i.e., a DVF which contains one motion vector per pixel, is quite high. Since the motion vectors are highly correlated, a predictive scheme should be used for encoding a dense DVF.

A class of motion estimators which are based on the spatio-temporal constraint equation and the prediction of the current motion vector, are pel-recursive motion estimators [170, 215]. The basic idea behind a pel-recursive motion estimator is to use the motion vectors of neighboring pixels to predict the current motion vector and then to iteratively change this estimate until the energy in the DFD at this pixel is minimized. It can be shown [202] that minimizing the DFD energy enforces the spatio-temporal constraint equation.

In [170] the best estimate for the previous motion vector is used as the prediction of the current motion vector and the minimization of the DFD energy is accomplished by a steepest descent algorithm. In [215] the step size of the steepest descent algorithm is adaptively changed, depending on the magnitude of the intensity gradient. The step size is inversely proportional to the magnitude of the intensity gradient which results in an accurate DVF estimation around edges and a fast algorithm in flat areas.

If the motion vectors used for the prediction of the current motion vector are selected in such a way that motion vectors which have already been transmitted are only used, then the correction term of the current motion vector needs only be quantized and sent. Sending the correction term can be achieved with a much smaller bit rate than the dense DVF since its entropy is much smaller. A recent scheme which is based on this concept is presented in [206, 11]

The main problem of pel-recursive motion estimators occurs in areas where there is a discontinuity in the motion field. Discontinuities appear mostly between moving objects and the stationary background. Since the prediction will fail in such areas, the steepest descent algorithm will not find a good local optimum and hence the DFD energy can be quite high. Ways to improve the accuracy of pel-recursive motion estimators are discussed in [41, 40] where spatially adaptive prediction is employed, and in [18, 19] where the motion discontinuities are modeled with the use of line elements. In [27] a pel-recursive motion estimators for quantum-limited image sequences is proposed and in [20] the general problem of pel-recursive motion estimation using noisy image sequences is reviewed.

2.2.2 DFD quantization and encoding

After the DVF has been estimated, the DFD can be calculated using Eq. (2.2). In other words, the predicted frame $\tilde{f}_k$ is formed by motion compensating the previously reconstructed frame $\hat{f}_{k-1}$ along the directions indicted by the DVF $\vec{d}_k(\vec{r})$. Then the predicted frame is subtracted from the current frame to form the DFD which needs to be quantized and encoded. The quantization operation is labeled "Q" in Fig. 2.2. Note that the encoding of the quantized DFD is not shown in Fig. 2.2, since this operation is lossless and can therefore be inverted. By not showing this operation, the decoder can be easily identified within the encoder structure.

PCM, DPCM and Vector quantization

The simplest way to quantize and encode the DFD signal is pulse code modulation (PCM). In PCM, the value of each pixel is quantized and encoded independently of the values of the neighboring pixels. When there is significant correlation between neighboring pixels, differential PCM (DPCM) is appropriate. In DPCM, the value of the current pixel is predicted using the neighboring pixels and the difference between the prediction and the actual value is quantized and encoded.

If the motion estimation was successful, there should be little correlation left in the DFD and hence PCM is usually preferred over DPCM. The main problem in PCM is the design of the scalar quantizer. If the number of reconstruction levels is given, and the index of the reconstruction level is sent using a fixed length code word, then the optimal quantizer for a certain source statistic is the well known Max-Lloyd quantizer [109, 118]. The Max-Lloyd quantizer is optimal in the sense that the resulting mean squared error is minimized.

Sometimes it is more convenient to hold the entropy of the quantized output constant while minimizing the resulting distortion. These quantizers are called entropy constrained quantizers [228]. For a general distribution it is quite hard to find the optimal entropy constrained quantizer. In [197] a fast method is presented for the optimal entropy constrained scalar quantization of exponential and Laplacian random variables. The methods employed to find these quantizers are quite similar to the ones we present in chapter 3. Clearly entropy constrained quantizers are better suited when the output of the quantizer is entropy coded and the Max-Lloyd quantizer is a better solution for fixed length codewords.

Some of the early image compression schemes employ PCM or DPCM [125], but the compression ratio for reasonable distortions is quite small. One example of a recent video compression scheme which employs PCM encoding of the DFD is presented in [123, 38]. The main reason for this is that the authors wish to derive a formula for the entropy of the quantized DFD. Since this is a hard problem in general, the authors select a PCM scheme with a uniform quantizer and assume a Laplacian distribution for the DFD, in achieving their goal.

Instead of quantizing each pixel independently, pixels can be grouped into a vector and then quantized jointly. This general idea is called vector quantization (VQ) [61]. VQ has become quite popular for image and video coding [131], but it is mostly applied in combination with a transform coder or a sub-band coding scheme. Nevertheless VQ can be applied directly to the DFD. This is most easily achieved by dividing the DFD image into equally sized squares of side length s. These squares can be represented by a single column vector of dimension $N = s^2$ by stacking the columns of the block. Each of these blocks represents a point in $\mathcal{R}^N$.

The idea behind VQ is to partition $\mathcal{R}^N$ into M regions, called Voronoi regions, where each region is represented by its centroid r_i. The collection of the centroids is then called the vector code book $C = \{r_1, \ldots, r_M\}$. The quantization works as follows: the distortion between the input vector and each of the code book vectors is calculated and the code book vector which results in the smallest distortion is selected as the quantized value. Since the code book is known at the encoder and the decoder, only the index of the code book vector needs to be transmitted. The decoding procedure is very simple since the decoder just selects the code book vector corresponding to the transmitted index.

VQ works best when the points in $\mathcal{R}^N$ are distributed in clusters. When the distribution of the points in $\mathcal{R}^N$ is completely uniform, VQ will not perform better than scalar quantization. The conditions necessary for the points to cluster are that there exists high correlation within the vectors belonging to the same Voronoi region, and the correlation between vectors which do not belong to the same Voronoi region is low. It is often the case that VQ is not directly applied to the signal, but some signal processing is performed first. The goal of that processing is to obtain a constellation as described above. In [102] a method is presented which tries to accomplish this with vector processing.

The main advantage of VQ over scalar quantization stems from the fact that VQ can utilize the correlation between pixels. Note that this correlation does not have to be a linear correlation as measured by the correlation coefficient, but can be of nonlinear nature. In fact, when there is nonlinear correlation among pixels, VQ can outperform scalar quantization even when the correlation coefficient is zero [103]. The reason for this is that the Voronoi regions of VQ can be of almost arbitrary shapes, whereas the shapes implied by scalar quantization are always rectangular.

As with scalar quantization, there exists the problem of finding the reconstruction levels, or in the VQ terminology, the code book vectors. Again, if we assume that the code book indices are encoded using a fixed code word length, the classical algorithm to find a code book which minimizes the distortion is the generalized Lloyd-Max algorithm proposed by Linde, Buzo and Gray [106], which is often called the LGB algorithm. The LGB algorithm is an iterative clustering scheme which is guaranteed to converge to a local optimum and requires training data which should be a representative statistical sample.

In the classical VQ scheme an exhaustive search is necessary to find the best match between the code book vector and the input vector. In order to improve the speed, a tree structure can be added to the code book, resulting in a tree searched VQ (TSVQ) [69]. The disadvantage of TSVQ is that its code book requires more storage and the code book vector selected during encoding might not be the one with the smallest distortion.

Quad tree (QT) coding

We discussed the QT structure in section 2.3, where it was used to efficiently represent an inhomogeneous DVF. By inspecting the DFD it is apparent that there are large homogeneous regions where the motion estimation was very successful. In those regions, the DFD energy is almost zero. Motion estimation usually fails around the boundaries of moving objects, which results in high energy in the DFD along these boundaries. In other words, the DFD appears like a line drawing of the boundaries of moving objects [234, 195]. Since QT-based techniques have been successfully applied to the encoding of binary line structured images [33], it seems natural to encode the DFD using a QT structure, and hence there are several different video coding schemes which are based on a QT encoding of the DFD [234, 195].

Transform coding

Transform coding has proven to be a very powerful method for the encoding of the DFD. In transform coding, the DFD is transformed (usually block wise) into another domain, where the transform is linear and invertible. The advantages of transform coding are twofold. First it is easier to measure the perceptual importance in the transform domain and second the transform is designed to de-correlate the DFD and to compact its energy into a few coefficients.

In a statistical sense, the optimal transform for the complete de-correlation of the data is the Karhunen-Loève transform (KLT) [61]. Note that for the discrete case, the KLT is sometimes called the Hotelling transform [86]. Also the discrete KLT is very closely related to the singular value decomposition [194] and the principal component analysis [68]. Besides completely de-correlating the data, the KLT is also optimal in the sense that the fewest number of coefficients are necessary to achieve a certain mean squared error [171]. This clearly shows the energy compaction property of the KLT.

The main drawback of the KLT is that the basis functions are image dependent, and therefore they change from block to block and from frame to frame. This is a serious drawback for compression since the decoder needs to know these basis functions, but transmitting them is not an option since this would result in a very high bit rate.

DCT-based coding

The discrete cosine transform (DCT) [166] is the most commonly used transform for image and video coding purposes. In fact, it has been so successful that every existing video coding standard is based on the DCT.

There are several different forms of the DCT [115]. The one used in the standards is the DCT-II which is of the following form,

$$F(u,v) = \frac{2}{N}C(u)C(v)\sum_{m=0}^{N-1}\sum_{n=0}^{N-1} f(m,n)cos(\frac{\pi(2m+1)u}{2N})cos(\frac{\pi(2n+1)v}{2N}), \tag{2.8}$$

where $F(u,v)$ is the coefficient in the transform domain at location (u,v), $0 \leq u \leq N-1, 0 \leq v \leq N-1$, and $f(m,n)$ is the coefficient in the original domain at location (m,n), $0 \leq m \leq N-1, 0 \leq n \leq N-1$. The function $C(x)$ is defined to be one except when x is equal to zero, in which case $C(0) = 1/\sqrt{2}$. This definition of the scaling function $C(\cdot)$ results in a unitary transform. Unitary transforms are energy preserving (L_2 norm preserving) which means that the following relationship holds,

$$\sum_{u=0}^{N-1}\sum_{v=0}^{N-1} |F(u,v)|^2 = \sum_{m=0}^{N-1}\sum_{n=0}^{N-1} |f(m,n)|^2. \tag{2.9}$$

Note that we will use this property later on to calculate the mean squared error efficiently in the DCT domain.

The popularity of the DCT is based on several different properties. Most importantly the basis functions of the DCT are image independent, unlike the KLT. It is shown in [5, 30] that the DCT approximates the KLT for a first order Gauss-Markov process when the correlation coefficient approaches ones. In practice, for natural images, the DCT approximates the de-correlation and energy compaction property of the KLT quite well. Another advantage of the DCT is that fast algorithms are available for its calculation. Some of these algorithms are direct methods whereas a popular class of algorithms is based on a fast Fourier transform (FFT) implementation [213]. It is well known [140] that the discrete Fourier transform (DFT) of a signal $\{x[n]\}_{n=0}^{N-1}$ is the Fourier series of the periodic extension of the signal. Hence if the beginning of the signal is significantly different from its end, a jump occurs at the boundary and the DFT contains high frequency energy due to this jump. In addition, even if the finite signal is real, the DFT is in general complex. The possible energy at high frequencies and the complex nature of the DFT results in the fact that the DFT is not commonly used for compression. It can be shown [103] that the DCT is related to the DFT. If the original signal is mirrored around its last element, $y[n] = x[n]$, $0 \leq n \leq N-1$, and $y[n] = x[2*N-1-n]$, $N \leq n \leq 2*N-1$, a signal $y[n]$ of twice the length of $x[n]$ is created. The first N samples of the DFT of $y[n]$ can directly be related to the DCT of $x[n]$ and this relationship is the basis of the fast DCT algorithms which utilize the FFT. This relationship also clarifies why the DCT does not suffer from the DFT problems, since the signal is first mirrored before it is periodically extended and the DFT is calculated. The mirroring leads to a smooth extension and hence there is no introduction of high frequency energy because of the periodic extension. In addition, the mirroring results in an even sequence and hence the DFT and the DCT are real. For the most commonly used block size, which is 8×8, there are VLSI implementations of the DCT available [15] and higher order DCTs, such as 16×16 can be recursively computed using an 8×8 DCT [77]. The availability of VLSI implementations of the DCT make it very attractive for a hardware-based real time implementation.

After the image block has been mapped into the DCT domain, most of the energy is represented by the low frequency coefficients. This is the idea behind energy compaction. For lossy coding, these coefficients need to be quantized. The most common scheme is to scalar quantize each coefficient independently using a uniform quantizer with or without dead-zone. For JPEG, a uniform mid-tread quantizer is used for each DCT coefficient. However, a quantizer with a different step size is used for each coefficient. The lower frequencies are quantized more finely whereas a coarser quantization is used for the higher frequencies. This weighting of the lower frequencies is an attempt to incorporate some properties of the human visual system into the coding scheme. JPEG can achieve compression ratios of up to a factor of 10 without noticeable degradation in the subjective image quality. After quantization, most of the transform coefficients are zero and the coefficients are scanned in a predetermined order. Usually a zig-zag scan (see Fig. 2.4) is used, where this scanning order attempts to order the coefficients in increasing order of spatial frequency. The resulting

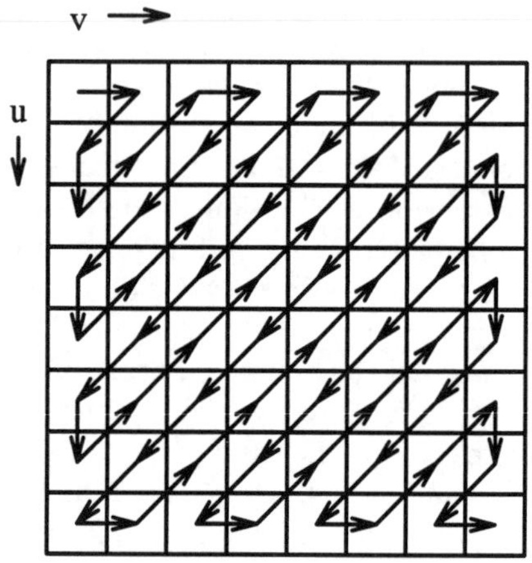

Figure 2.4 Zig-zag scan for the DCT coefficients.

string is then run-length and entropy coded which results in a very compact representation of the quantized transform coefficients. Run-length encoding works as follows: the string is searched for non-zero coefficients and their position is encoded by recording the run of zeros before this coefficient was found and the value of this coefficient. Hence each non-zero coefficient results in a two dimensional symbol (run,value), which then is entropy coded. If there are no more non-zero coefficients left, an end of block (EOB) symbol is used. Instead of scalar quantization, vector quantization could also be used for the DCT coefficients. All existing standards however, which are based on the DCT, use the scheme outlined above.

Besides its many advantages, DCT-based DFD encoding has some problems too. The biggest of them is the fact that DCT is a block-based transform and at very low bit rates the block boundaries start to become visible [160], since each block is quantized independently. There have been some efforts to reduce these blocking artifacts. A straightforward approach is to let the blocks overlap and then average the pixels which belong to two blocks. This will clearly reduce the blocking artifacts but on the other hand, the compression is not as efficient since there are pixels which belong to more than one block and hence they get encoded more than once. One solution to this problem is the lapped orthogonal transform (LOT) [113, 112] where the overlapping of blocks is achieved without adding additional redundancy.

Another approach for removing blocking artifacts is based on post-processing of the decoded frames. Early work in this field is reported in [161, 167]. In a recent approach [147], a spatially adaptive maximum a posteriori (MAP) solution was derived, according to which the sharpness of the existing edges in the image was preserved, while the artificial edges at the block boundaries were smoothed over. Another recent approach uses projection onto convex sets (POCS) in also deriving both spatially invariant [231] and spatially adaptive [232] solutions.

It is interesting to notice that the DCT is an approximation to the KLT when the correlation between neighboring pixels is very high. This is certainly the case for a natural image, but in the case of the DFD, such a high correlation should not be expected if the motion estimation was successful. In [195] the correlation coefficient of the DFD was measured to be between 0.3 and 0.5, which is much lower than for a regular image. Therefore the DCT does not approximate the KLT when applied to the DFD frame in general. Still the ability to quantize the low frequencies more finely than the high frequencies and the very efficient run-length/entropy coding of the remaining coefficients make the DCT a popular choice for the DFD encoding.

Sub-band and wavelet coding

We have concentrated our discussion of transform coding of the DFD on the DCT since this is the most commonly used scheme. In recent years, two methods, sub-band and wavelet coding, have become popular for image compression [229]. The basic idea behind sub-band coding (SBC) [211] is to split the spectrum of the image into different sub-bands, subsample these sub-bands, and encode each sub-band separately using a coder and bit rate closely matched to the statistics of that particular band. Results on SBC of images were first reported in [230], where adaptive DPCM is used to encode the individual sub-bands. Other work in SBC of images show the use of both DPCM and PCM coders [64, 221], while results of incorporating vector quantization in a SBC scheme are presented in [223, 220]. Even though these schemes have proven to be very powerful for image compression [229], they are not very often used to encode the DFD. The main reason for this is that the motion estimation is usually accomplished using block matching. Therefore the DFD will have a block structured nature and hence a block-based transform such as the DCT is a logical extension. For SBC schemes, which are not block-based, but decompose the entire DFD, this block structure can result in high frequency energy, which is hard to encode efficiently. In [62] an attempt is made to fit sub-band coding into an MC coder and we discuss the more common approach of three dimensional sub-band coding in section 2.3.2.

Wavelet coding (WC) relates to SBC in that wavelet theory [212] can be used to derive filter banks for the sub-band decomposition. The sub-band decomposition is usually accomplished by perfect reconstruction quadrature mirror filters (QMF) and the resulting multi-resolution decomposition can be viewed as a signal expansion onto a set of wavelet basis functions. When the frequency bands are spaced in octaves, the output of the QMFs are the discrete time wavelet series expansion of the input signal. Note that wavelet theory has a very wide application and is not restricted to signal compression alone.

2.2.3 Generic MC-DCT scheme

In this section we present a generic motion compensated (MC) scheme which is the basis for all video coding standards, from H.263, the very low bit rate standard ($\leq$ 64 kbits/second) to H.261, the current video conferencing standard at $p \times 64$ kbits/second ($6 \leq p \leq 30$) for integrated services digital network (ISDN) [107], to MPEG-1, the multimedia standard mostly intended for CD-ROMs ($\leq$ 1.5 Mbits/second), to MPEG-2, intended for digital TV (2-20 Mbits/second).

All these schemes are block-based, where the common block sizes are 8×8 and 16×16. This means that the frame is first segmented into blocks of the same size and then each block is encoded independently. For each block one motion vector is assigned and commonly block matching is used for this, although the standards do not specify how this motion vector is found. The DVF is encoded using DPCM and entropy coding whereas the resulting DFD is encoded using the DCT and the quantization and entropy coding scheme explained in section 2.2.2. Note that the bit rate range for which this scheme has been successfully applied covers three orders of magnitudes, which is quite astonishing.

Consider Fig. 2.5 which displays the generic MC-DCT coder. The original frame f_k, and the previously reconstructed frame $\hat{f}_{k-1}$ are used to find the best DVF, $\vec{d}_k$. As mentioned above, this is usually accomplished by block matching (BM). Because of the high spatial correlation, DPCM is applied to the DVF and the resulting prediction error is entropy coded. The DVF and the previously reconstructed frame are then used to construct the predicted frame $\tilde{f}_k$, which is then subtracted from the original frame to form the DFD. The DCT is applied to the DFD and the DCT coefficients are quantized, run-length and entropy coded. The DCT needs then to be inverted (IDCT) and when the

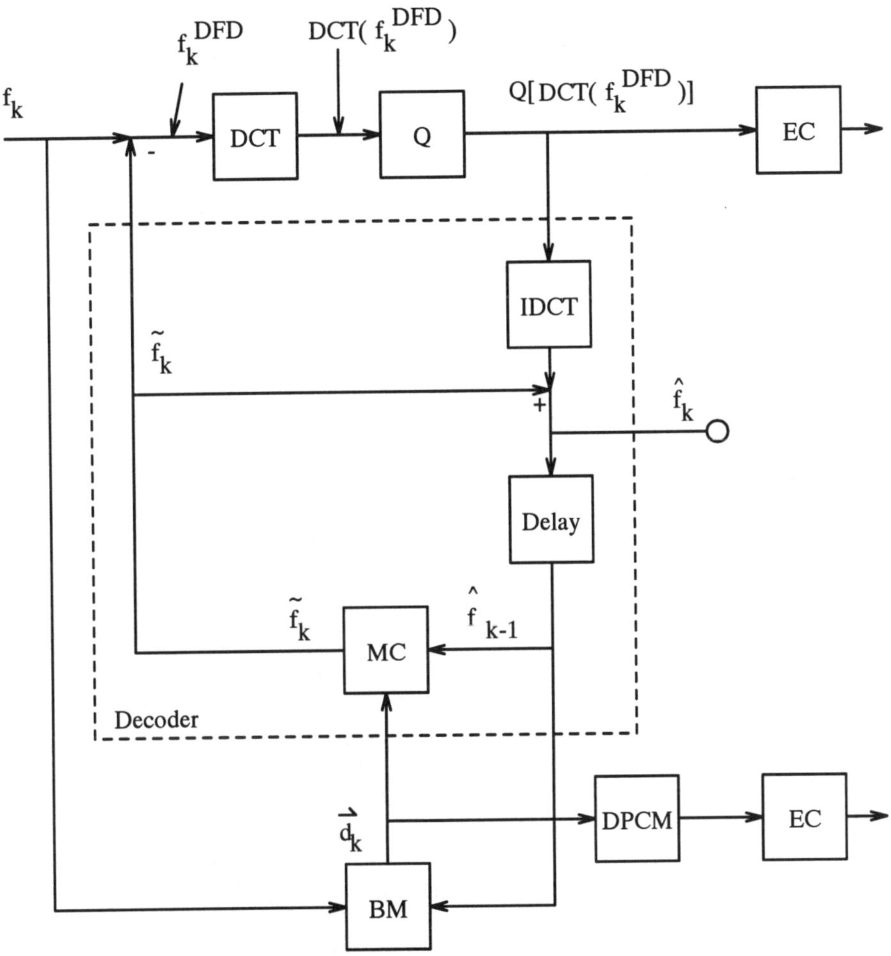

Figure 2.5 Block diagram of a generic MC-DCT coder. DCT: discrete cosine transform; IDCT: inverse DCT; Q: quantizer; EC: entropy coder; Delay: frame delay; MC: motion compensation; BM: block matching; DPCM: differential pulse code modulation.

predicted frame is added to this quantized DCT, the reconstructed frame is generated.

2.3 THREE DIMENSIONAL WAVEFORM CODING

In this section we discuss some schemes in which motion compensation is not explicitly performed in the spatial domain, but the three-dimensional signal statistics are used for the compression of a given space-time volume. The last scheme, pyramid coding, does contain explicit motion compensation in the spatial domain, but in a different fashion than presented in the previous section.

2.3.1 Three dimensional transform coding

Three-dimensional transform coding (3DTC) is a straightforward extension of two-dimensional transform coding in that a cube (x, y, and t) instead of a block (x and y) is transformed. Clearly for a stationary cube, most of the energy in the temporal direction is concentrated around the DC component and hence the use of similar quantization techniques in the temporal direction as in the spatial dimensions results in an efficient representation of the video sequence. If there is motion in the cube, then the higher frequencies have to be carefully quantized since otherwise blurring will result. Notice that there is an inherent delay in this scheme, since j (for a cube of temporal length j) frames have to be stored first, before the transform can be performed. The main advantage of 3DTC is that no motion estimation step is required. Overall 3DTC has not been proven successful since it can be easily outperformed by the MC-DCT scheme discussed above. This is mainly due to the fact that motion is a local phenomenon in the spatial domain and it can therefore be compactly represented by, for example, a motion vector.

2.3.2 Three dimensional sub-band coding

In three dimensional sub-band coding (3DSBC) [90, 214], the idea is to split the three dimensional spectrum into different, three dimensional sub-bands, down-sample these bands and encode them with quantizers and rates matched to the statistics of the particular band. There are three main advantage of

3DSBC over 3DTC. First, since the video sequence is not segmented into blocks, no blocking artifacts will occur. Second, different filters can be applied to the spatial and temporal dimensions, and third, sub-band coding is spatially and temporally scalable. In general the temporal filter should be fairly short, otherwise the lowest frequency band will show blurring in areas where there is motion and the high frequency bands will have energy pockets at the location of the motion. Short filters also result in less delay and the requirement for frame storage is smaller. Since sub-band decomposition results in a multi-resolution representation of the sequence, the lowest band is a low-passed and down-sampled version of the original sequence and the other bands carry information about higher frequencies and higher frame rates, hence the resulting video coding scheme is scalable, both in time and space. If we compare the 3DSBC and the MC-DCT approaches the main advantages of 3DSBC are: spatial and temporal scalability, no blocking artifacts, simplicity of implementation and no motion estimation step. The main disadvantage is that so far the MC-DCT approach can outperform 3DSBC in terms of image quality for the same bit rate.

If we are willing to drop the advantage of not having to perform motion estimation, then it seems more natural to perform the temporal filtering along the motion trajectories. Indeed there have been approaches proposed based on this idea [138], but it is quite hard to include motion estimation into the different sub-bands, especially for higher frequency bands [212]. A more natural combination of motion compensation and multi-resolution representation is presented in the next section.

2.3.3 Pyramid coding

Originally pyramid coding was introduced as an image coding scheme [25]. The basic idea behind pyramid coding is to encode a low-pass filtered and down-sampled version of the original image, and use an interpolation scheme to estimate the original image from this low resolution version. Then the prediction error is build as the difference between the estimate and the original image. Hence the image has been decomposed into two parts, a coarse (low resolution) version and a prediction error. The better the prediction of the original image from the coarse image, the more efficient the method, since the coarse image and the prediction error need to be quantized and encoded. Clearly this concept can be iterated, or in other words, one can again decompose the coarse image into an even coarser one and a new prediction error. This pyramid structure clearly represents a multi-resolution representation of the original image and it

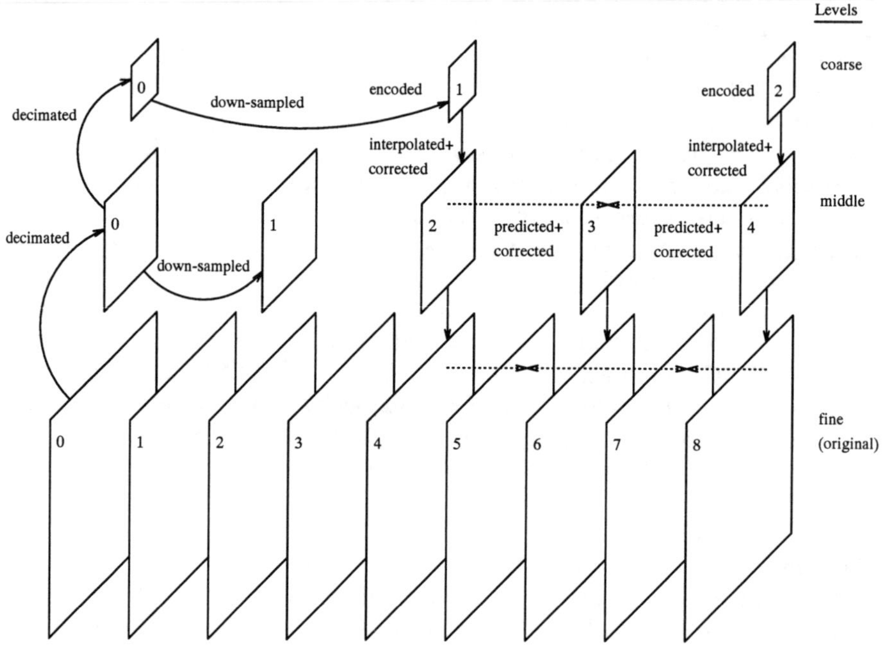

Figure 2.6 Three level pyramid video coding. The spatial levels are achieved by decimating by a factor of two and the temporal levels are obtained by down-sampling by a factor of two. The coarse levels are encoded independently. They are then used to interpolate the middle levels and the interpolation is corrected using the encoded interpolation error. The temporal interpolation is achieved by motion compensated prediction, where again the prediction is corrected using the encoded prediction error.

is very closely related to wavelet and sub-band coding. The main difference is that the decimation and interpolation schemes are not such that the resulting signals can be critically down-sampled without loss of information. Therefore there is an inherent signal expansion involved in this scheme, which seems contradictory to the aim of video compression. It can be shown [212] that this expansion is fairly small for the three-dimensional case.

The pyramid video coding scheme can be explained best by means of an example, which is taken from [212]. Consider Fig. 2.6 which shows a three level pyramid decomposition of an original video sequence which is the finest level of the pyramid. The different spatial levels are achieved by decimating by a factor of two and the temporal levels are generated by simply frame skipping. The coarse levels are encoded independently and they represent the lowest quality approximation of the original video sequence. If a higher spatial resolution is desired, the coarse levels can be used to interpolate the middle levels, and this interpolation is then corrected using the encoded interpolation error. To increase the temporal resolution, i.e., the frame rate, bi-directional motion compensated interpolation is used, where again the encoded prediction error is used to correct the prediction. To generate the finest level, the interpolation-correction and prediction-correction operations are repeated.

2.4 MODEL-BASED VIDEO CODING

In this section we present schemes which employ complex scene models to represent the motion in the scene more accurately. The only scene model discussed so far was that of blocks moving translationally into different directions. This is the model used by all current video coding standards. It has been quite successful, mostly due to its simplicity and robustness. In utilizing more complex scene models, the goal is to more accurately predict the current frame than when block matching is used. If the prediction is better, then eventually, the DFD does not have to be sent at all, or only in areas where the prediction failed completely due to uncovered background or the appearance of a new object.

In model-based coding, an attempt is made to explicitly or implicitly reconstruct the real three-dimensional world from the two-dimensional projections [101]. By having an accurate mathematical model of the real world, for analyzing and synthesizing image sequences, very high compression is conceptually possible. Basing the prediction of the current frame on such a model might still be inaccurate, but if the model is close enough to the real world, a human observer might not be able to recognize the inaccuracies since they still reflect a physically possible scene.

2.4.1 Object-oriented coding

Object-oriented coding (OOC) was originally proposed in [126] and there have been many improvements and changes to the basic scheme since then [73, 144, 74, 75, 37, 124, 145, 60, 146]. The basic idea behind OOC is to segment a given frame into different objects conforming to the object model employed. Each object is defined by three properties, motion, shape and color information (in this section, color represents luminance and chrominance). OOC is an analysis-synthesis-based scheme which means that the current frame is analyzed using the object models, with the goal of extracting the parameters of the objects and then the reconstructed frame is synthesized using these parameters and the object models. The difference between the original frame and the synthesized frame is then used to improve the parameter estimation in the analysis stage.

The analysis stage is by far the most challenging part of an OOC scheme and approaches for finding the object parameters and the frame segmentation are reported in [92, 36, 76]. The reason why the analysis stage is complicated lies in the fact that for a reliable motion estimation, the segmentation of the frame into different objects needs to be available. Unfortunately, the objects are defined by their parameters, one of which is the motion, which implies that the motion needs to be known before the frame can reliably be segmented. Hence we have the problem that the segmentation needs to be known to find the motion and the motion needs to be known to find the segmentation. Clearly an iterative approach is an obvious solution to this problem, though it is quite hard to prove convergence for such iterative schemes.

There are different models which can be employed to describe the scene. In [126] 2-D and 3-D rigid objects with 3-D motion are studied. This is extended in [73, 74] to 2-D flexible objects with 2-D motion and in [145] to 3-D flexible objects with 3-D motion. The reason for introducing more and more complicated models is twofold. First the real world is not made up of rigid 2-D objects and hence more complex models should result in a better representation of the real world. Second, a higher coding gain can be achieved when the prediction is improved, since the color information has to be sent in the so called "model failure" (MF) regions. MF regions are regions in which the prediction of the current frame, using a given scene model, is not satisfactory. Note that in contrast to a traditional waveform coding scheme, the DFD is not transmitted in OOC, even though the DFD might contain significant energy. The only color information transmitted is in the MF regions and this is usually accomplished by one of the established waveform coding schemes. Using the "rigid 3-D object with 3-D motion" model, the MF regions of the "Miss America" sequence are about 9% of the total image [126] which can be reduced to only 4% when "flexible 2-D models with 2-D motion" are used [73]. An even smaller MF region percentage of 3% is reported in [145], where a "flexible 3-D objects with 3-D motion" model is proposed. The video compression scheme proposed in chapter 7 compares favorably to these numbers since it only transmits color information in about 6% of the frame. It achieves this low percentage because the employed motion estimation is very successful even though a simple motion model is used.

A major advantage of OOC is that the distortions resulting from the quantization of the motion and shape parameters are less objectionable than the distortions usually associated with waveform coders. In fact, a relatively large error is permitted in the shape representation before the distortion starts to be visible. This kind of distortion is referred to as "geometric distortion". In chapter 8 we present an segmentation encoding scheme which offers full control over the amount of geometric distortion introduced.

A main disadvantage of OOC is the complicated analysis part. The iterative algorithms employed tend to be non-robust and hence their performance has not been proven over a wide variety of video sequences. The problem of finding the motion and the segmentation of a frame is a very hard task when these advanced scene models are used. As mentioned above, iterative schemes are employed which first find the motion and then use the motion to find the segmentation and then iterate this procedure until some kind of a convergence criterion is satisfied. Ideally one would like an algorithm which simultaneously finds the segmentation and the motion. In chapter 7 we present an algorithm which finds the segmentation, motion and residual error simultaneously as the solution to a well defined optimization problem. In fact the resulting segmentation, motion and residual error are optimal in the rate distortion sense. The presented algorithm is guaranteed to find the solution and hence the scheme is extremely robust. In fact the entire algorithm works without a single threshold.

2.4.2 Knowledge-based and semantic coding

In knowledge-based coding, [51, 150, 7, 6, 28, 8] the encoder and decoder share some explicit knowledge about the structure of the frame. This is the main difference to OOC, where the structure of the scene is unknown and has to be estimated. Usually this structural knowledge is represented by a wire-frame model of the objects in the scene. Knowledge-based coding is typically applied to video phone sequences, and hence most wire-frame models used for video coding are head models, such as the ones presented in [148, 150, 172, 7].

The main aim of knowledge-based coding is the very low bit rate coding of talking heads. An important attribute when transmitting facial information is the facial details which human beings find necessary when communicating. There are two broad categories to be considered. The first is conformation, controlling the structure of an individual face, and the second is expression, controlling the emotional information. Included in conformation are skin color, aspect ratio of the face, and modifications to nearly every aspect of the facial structure. Included in expression are the attributes of the eye, eyelids, eyebrows, jaw rotation, width of the mouth and positions of the lips and corners of the mouth. Usually the hair and the shoulders are reproduced using two-dimensional texture mapping parallel to the image plane.

The wire-frame model used in [7] consists of approximately 400 triangles. The MBASIC scheme proposed in [7] first adapts the generic head model to a specific user. Then the global motion of the head is estimated using the wire-frame model. The estimated global motion is used to motion compensate the global motion and the local motion is estimated using semantic coding.

The idea of semantic coding for facial expression is to decompose the local motion into a linear combination of so called action units (AU). AUs are the basic elements of the Facial Action Coding System (FACS) proposed in [42]. The proposition is that every facial expression can be decomposed into a linear combination of AUs and since the AUs are known at the encoder and the decoder, this is a very efficient representation of the local motion.

Knowledge-based and semantic coding can achieve extremely low bit rates, since the scene is encoded at a very high structural and semantic level. Clearly the approach is limited in the sense that it can only handle object for which the system has been designed, such as talking heads for the video phone example. Besides very low bit rate coding, knowledge-based concepts are also very useful for animating synthetic talking actors. In other words, one would like to achieve a realistic animation of a speaking person, using only a model of the person and a script of what he or she is supposed to say.

There has been considerable research in the area of speech driven facial animation. The two driving forces are computer animation [111, 88, 100, 148, 149, 150, 151, 71, 152, 219, 218, 203, 204, 205, 127], and model-based video coding of speaking facial images [121, 122]. Also psychologists and speech and hearing impairment researchers [117, 116, 31, 32, 22, 43, 17, 23] use synthesized audible and visual speech to perform audio-visual perception experiments. The field of speech driven facial animation can be separated into two and three dimensional approaches (2D and 3D), where the 2D versions have the advantage of speed but the 3D implementations give the animator more degrees of freedom.

Many of the early works were based on a 2D representation of the lip shapes. In [44] Lissajou figures were generated on an oscilloscope to simulate lip movements. In [120] one of the first models for the coarticulation effects of lip shapes was proposed. The lips were displayed on a vector display device and at about 4 times slower than real time. In [23] for the first time a simple face was displayed in real time using a vector display. Current research uses mainly 3D models because of the additional degrees of freedom these models offer. This comes with a high price though. Animating such a model is computationally expensive and so far we know of only one case where the animation of a complex model is done in real time [122] using high speed equipment.

There are two trends in the field of 3D facial animation which are best represented by the work of their respective inventors, Parke and Waters. Parke started the interest in 3D facial modeling with his Ph.D. dissertation in 1974 [148]. Parke's model is a parametrically controlled polygon topology (wire frame) where approximately 900 small surfaces arranged in 3D make up the face [100, 148, 149, 150, 151]. The animation of the face is done by controlling a set of about 50 parameters, where about 10 are used to control the movements involved in speech production. In his dissertation, Parke produced sentences by observing his own articulation and simulating it, using his facial model. In [100] Parke and Lewis automated the animation of speech by using linear predictive coding to analyze recorded speech and recognize the vowel sounds which are then used to animate the facial model. Parke's software is the basis for the work in [152, 71]. An existing speech synthesizer by rule was used and the parameters for the facial animation were added to the list of actions to take for a certain phoneme. The program accepts text and then parses it into phonetic units which then drive a speech synthesizer and the facial animation system. In [117, 116, 31, 32] Parke's model was also used including some of the additions made in [152, 71]. The main improvement was the incorporation of the tongue and the extensive study of coarticulation effects which where integrated into the animation of the face [32].

In contrast to Parke's work, Walters uses the underlying anatomical structure of the human face to construct his 3 layer model [219, 218, 203, 204, 205]. He use an approximation of the skull and the pivoting jaw as the first layer. Then muscle tissues and their insertions are placed over the skull. This requires complex elastic models for the compressible tissues. A covering surface layer changes according to the underlying structures and animation is done by providing information about a set of contraction and relaxation of certain muscles. In [156] a similar approach together with the FACS is employed. More recently, linguistic knowledge has been used to incorporate coarticulation [153].

The above mentioned approaches are the dominant ones, but there are other facial animation systems which are not based on Parke's or Walter's models. In [127] a B-spline surface is used to model the face. B-spline control parameters were obtained for each phoneme and the face is animated by concatenating the phonemes to achieve the desired speech sequence. In [233] face is animated by using texture mapping techniques, where video images are projected onto a 3D wire frame model. This method is aimed at overcoming the problems with facial animation systems which generate images that look more like cartoon characters than humans.

2.5 SUMMARY

In this chapter we reviewed the most popular lossy video compression schemes. We first pointed out the difference between lossy and lossless schemes and then presented the paradigm of motion compensated waveform coding. We reviewed three dimensional waveform coding and discussed the three most common approaches. Finally we presented model-based video coding schemes, specifically object-oriented approaches and knowledge-based schemes.

3

BACKGROUND

In this chapter we introduce the background necessary for the remainder of the monograph. This chapter is organized as follows: In section 3.1 we discuss rate distortion theory which is mostly concerned with the derivation of absolute bounds on the performance of a lossy data compression scheme. In section 3.2 we compare rate distortion theory to operational rate distortion theory, which is concerned with the optimality of a particular scheme. In section 3.3 we present the Lagrangian multiplier method in the context of operational rate distortion theory. In section 3.4 we give an overview of dynamic programming which is a general tool for sequential decision analysis. In section 3.5 we discuss a shortest path algorithm for a weighted directed acyclic graph which is related to dynamic programming and in section 3.6 we summarize this chapter.

3.1 RATE DISTORTION THEORY

In this section we introduce some basic concepts of rate distortion theory (RDT). We do not attempt to be complete or rigorous since RDT is quite a complex and mature field. The standard reference for RDT is [13]. RDT is one part of information theory and as such originated in the original paper by Shannon [188] and was extended by Shannon in [189]. A review of RDT with respect to image coding can be found in [35].

A central result of information theory states that the problem of source coding, i.e., "what information should be sent", and the problem of channel coding, i.e., "how should it be sent", can be separated without loss of optimality [13, 57].

RDT is concerned with the first problem, i.e., what information should be sent, and it therefore directly relates to lossy video compression.

The source coding theorem states that the entropy $H(X)$ of a source X is the minimum rate at which a source can be encoded without information loss. On the other hand, the channel coding theorem states that for error free transmission over a channel with capacity C, the rate of the source needs to be smaller or equal to C. We pointed out in the introduction that the main reason why video compression is a very important problem is due to the fact that video requires an enormous data rate for its representation which the available channels cannot accommodate.

The central entity in RDT is the rate distortion function (RDF) $R(D)$, which represents the lower bound on the rate necessary to represent a certain source with a given average distortion. This lower bound is with respect to any encoding scheme, or in other words, it implies that the block length used for the encoding tends to infinity. The RDF can be considered as an extension of the idea of entropy for the case when a certain distortion in the representation of the data is permissible.

Therefore, if a certain channel capacity C is given, the RDF can be used to find the necessary minimum average distortion D_{ave} so that the condition for error free transmission $R(D_{ave}) < C$ is achieved. Conversely the RDF can also be used to find the minimum capacity necessary to transmit a certain source with a given average distortion.

The following is a more quantitative introduction to some of the concepts in RDT and follows the presentation given in [35]. Consider Fig. 3.1 which shows the basic parts of a communication system. Since RDT is concerned with lossy source coding, we assume an error free channel encoder, channel, and channel decoder for the rest of this section. Suppose that the source is generating symbols which are discrete in amplitude and time. These symbols are encoded in blocks of length N, where each block is described by a message from a denumerable set of messages $\{X_i\}$ (since each of these blocks will be represented by a code word, this is a vector quantization scheme). We further assume that each of the messages occurs with probability $P(X_i)$. The entire system can now be mathematically described by the conditional probability $Q(Y_j|X_i)$ of message Y_j being delivered to the sink, given that the source produced message X_i. The marginal probability of the received message set is therefore,

$$T(Y_i) = \sum_i P(X_i)Q(Y_i|X_i). \tag{3.1}$$

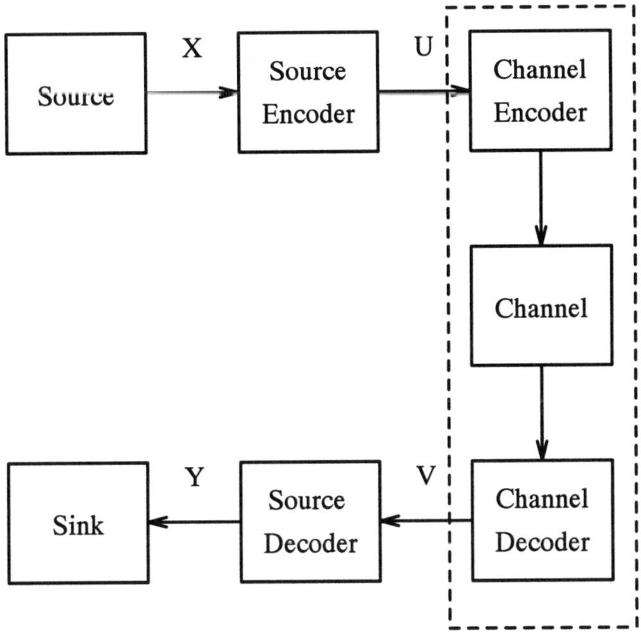

Figure 3.1 Communication system.

The average mutual information for a block of length N, $I_N(X,Y)$ is the information theoretical measure of the information transmitted from the source to the sink and is defined as follows,

$$I_N(X,Y) = \sum_i \sum_j P(X_i)Q(Y_j|X_i)\log(\frac{Q(Y_j|X_i)}{T(Y_j)}). \tag{3.2}$$

The units for the information are based on the base of the log used; if 2 is used for the base, then the resulting information is measured in bits. This definition of mutual information is based on the concept of self information of a discrete random variable. The self information gained by observing the outcome a of a discrete random variable A is defined to be $\log(\frac{1}{Pr(A=a)})$, where $Pr(\cdot)$ is the probability measure. The intuition behind this definition is the fact that the self information is high for outcomes with little probability and low for outcomes with high probability. In other words, when we observe an event which is almost certain, such as the Chicago Bulls winning the NBA, the information gained by observing this event is very small. On the other hand if the event is very unlikely, such as the Chicago Cubs going to the World series, the information gained by observing this outcome is large. There are other reasons for selecting this measure of information, such as the additivity of information for joint events and the fact that information should always be non-negative [57].

The concept of self information can now be used to explain the definition of average mutual information in Eq. (3.2), which can be written in the following way,

$$I_N(X,Y) = E_{X,Y}\left[\log(\frac{1}{T(Y_j)}) - \log(\frac{1}{Q(Y_j|X_i)})\right], \tag{3.3}$$

where $E_{X,Y}[\cdot]$ is the expectation operator with respect to X and Y. The term inside the brackets in Eq. (3.3) is the difference between the self information gained by observing a message Y_j at the sink and the information gained by observing a message Y_j at the sink given that we know that message X_i was sent. In the case of error free encoding, clearly the second information is zero, since knowing X_i implies knowing Y_j. Error-less encoding implies that $Y = X$ and $T(Y_i) = P(Y_i)$. Hence the information transmitted is equal to the average self information of Y, which is also called the Nth-order entropy of Y, $H_N(Y)$, which again is equal to the Nth-order entropy of X, $H_N(X)$,

$$I_N(X,Y) = H_N(X) = H_N(Y) = \sum_i P(X_i)\log(\frac{1}{P(X_i)}). \tag{3.4}$$

Note that optimum error free coding requires at least $H_N(X)$ bits.

When source coding results in an error, which is the case we are more interested in, then there is some information to be gained by observing Y_j even though we already know X_i. In general the average mutual information can be written as follows,

$$I_N(X,Y) = H_N(X) - H_N(Y|X), \tag{3.5}$$

which directly follows from Eq. (3.3).

Since average mutual information and entropy are both non-negative quantities, it follows that the average mutual information is always smaller or equal to the entropy of the source, which agrees with the intuition that no information can be added by the encoding stage. Clearly information can be lost during the encoding and hence the sink will not receive exactly what has been sent by the source. We have seen that for error-less encoding the conditional entropy $H_N(Y|X)$ is zero since knowing X implies knowing Y. In the other extreme, where knowledge of X does not give any statistical information about Y, the conditional entropy $H_N(Y|X)$ is equal to the entropy of Y, $H_N(Y)$, and hence the average mutual information is zero, or in other words, no information is transmitted. We are most interested in the case where the encoding scheme introduces an error, so that there is a strong statistical dependency between X and Y, hence $H_N(Y|X)$ is neither equal to zero nor equal to $H_N(Y)$.

Let $d(X,Y)$ be the distortion between X and Y. The expected distortion for a given conditional probability assignment $Q(\cdot|\cdot)$ is therefore,

$$\begin{aligned} D(Q) &= \frac{1}{N}E_{X,Y}[d(X,Y)] && (3.6)\\ &= \frac{1}{N}\sum_i\sum_j P(X_i)Q(Y_j|X_i)d(X_i,Y_j). && (3.7)\end{aligned}$$

The N-block RDF ($R_N(D_*)$) is the minimum of the average mutual information of Eq. (3.2) per symbol, for $D(Q)$ of Eq. (3.6) less than some value D_*. Since the source is given, the minimization has to be taken over the conditional probability assignment $Q(\cdot|\cdot)$, hence the N-block RDF is defined as follows,

$$R_N(D_*) = \inf_{Q:D(Q)\leq D_*} \frac{1}{N}I_N(X,Y). \tag{3.8}$$

Finally the RDF is obtained by taking the limit as the block size N goes to infinity,

$$R(D_*) = \lim_{N\to\infty} R_N(D_*). \tag{3.9}$$

Therefore, the RDF is the limit as the block size goes to infinity of the average mutual information per source symbol, subject to the constraint that the average distortion is less than D_*, where the minimization is performed over all encoding schemes as described by the conditional probability assignment $Q(\cdot|\cdot)$. The RDF has many interesting characteristic. For example it is continuous, convex, differentiable and non-increasing.

The channel capacity C is defined in a similar fashion, in that it is the maximum average mutual information between the channel input U and the channel output V, (the channel encoder and decoder are included in the channel). The optimization is carried out over the conditional probability assignments $W(V_j|U_i)$ which describes the channel mathematically. Hence the channel capacity C is defined as follows,

$$C = \lim_{N\to\infty}\sup_W \frac{1}{N}I_N(U,V). \tag{3.10}$$

After having defined the RDF we can now state the source coding theorem for a discrete source with a fidelity criterion [189]. The source coding theorem, when distortion is permissible, states that if $C > R(D_*)$ then, by using large enough blocks, a source can be sent over a channel with capacity C with an average distortion less or equal to D_*. The converse of this theorem is also true, that is, if $C < R(D_*)$ then it is impossible to send that source over a channel with capacity C having an average distortion less or equal to D_*.

The power of RDT stems from the fact that absolute bounds can be found on the performance of a lossy data compression scheme. Unfortunately, most of the proofs are existence proofs, which means that they prove there exists a bound, but they do not provide a mechanism for constructing a scheme which achieves this bound. Another problem with RDT is that the RDF can only be derived explicitly for simple sources and simple distortion measures.

3.2 OPERATIONAL RATE DISTORTION THEORY

In this section we discuss the concept of operational rate distortion theory (ORDT) and compare it to rate distortion theory (RDT) introduced in the previous section. We also review the field of optimal bit allocation since it directly relates to ORDT.

ORDT is based on the fact that every lossy data compression scheme has only a finite set of admissible quantizers. Therefore there is only a finite number of possible rate distortion pairs for any given source. These pairs form the quantizer function (QF). In Fig. 3.2 the number of possible quantizers is 27, 10 of which result in rate distortion pairs which constitute the operational rate distortion function (ORDF). Let $Q = \{q_0, \ldots, q_{M-1}\}$ be the set of all admissible quantizes, where $M = |Q|$, the cardinality of Q. Let $R(q_i)$ be the rate for a particular quantizer and a particular source and $D(q_i)$ the corresponding distortion. Hence the QF can be defined formally as follows,

$$\mathrm{QF} = \{(R(q_i), D(q_i))\}_{i=0}^{M-1}, \tag{3.11}$$

which shows the misnomer since the QF is really a set.

Let the quantizers which result in the ORDF be defined as follows,

$$Q_{ORDF} = \{q : q \in Q, R(q) \geq R(p) \Rightarrow D(q) < D(p), \forall\, p \in Q\}, \tag{3.12}$$

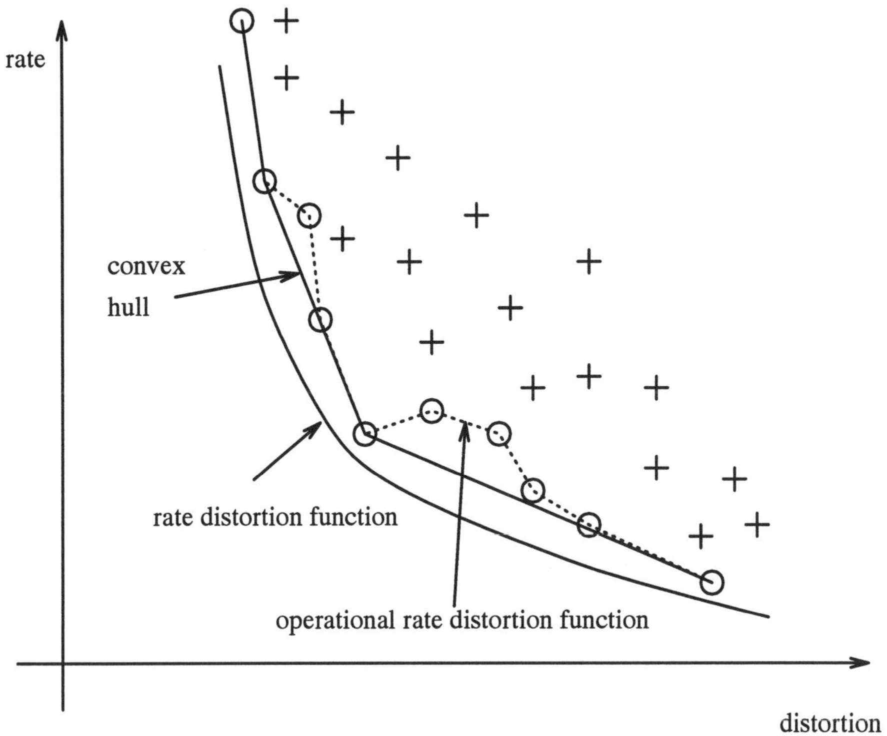

Figure 3.2 Operational rate distortion function. The crosses and circles represent all possible rate distortion pairs, which define the quantizer function, whereas the circles indicate the points of the operational rate distortion curve.

which is equivalent to the following definition,

$$Q_{ORDF} = \{q : q \in Q, D(q) \geq D(p) \Rightarrow R(q) < R(p), \forall\, p \in Q\}. \qquad (3.13)$$

The above two definition simply state that a quantizer can only belong to Q_{ORDF} if there is no other quantizer which results in a lower or equal rate and a lower distortion or equivalently, there is no other quantizer which results in a smaller distortion using the same or a smaller rate.

We relabel the quantizers in Q_{ORDF} from zero to $|Q_{ORDF}|-1$ such that $D(q_i) < D(q_{i+1})$, which by the definition of Q_{ORDF} implies that $R(q_i) > R(q_{i+1})$. Then the ORDF is defined as the following ordered set of rate distortion points,

$$\text{ORDF} = \{(R(q_i), D(q_i))\}_{i=0}^{|Q_{ORDF}|-1}, \tag{3.14}$$

which are indicated by circles in Fig. 3.2. Sometimes it is convenient to think of the ORDF as a continuous curve, and for that purpose we connect consecutive points in the rate distortion plot by a straight line, indicated by the dotted line in Fig. 3.2. Also note in Fig. 3.2 the lower half of the convex hull of the ORDF, which is very important for many bit allocation algorithms, and the RDF, which is a lower bound for the performance of the quantization scheme.

The idea of ORDT is to analyze a given compression scheme and to find its theoretical optimum performance, whereas RDT tries to find the absolute performance bound with respect to any scheme. While RDT is useful in assessing how close the performance of an actual scheme comes to the theoretical optimum, ORDT can be used to optimize the actual scheme to perform at its best. Performing at its best means in ORDT, that for a given rate, the quantizers are employed such that the resulting distortion is minimized. Hence ORDT is closely related to optimal bit allocation.

3.2.1 Optimal bit allocation

In optimal bit allocation the aim is to distribute the available bit budget among different sources of information such that the resulting overall distortion is minimized. Each of the information sources has usually a set of possible quantizers associated with it and hence the source and the quantizers can be described by a QF.

One of the first reports on optimal bit allocation is found in [79], where the problem of block quantizing (vector quantizing) correlated Gaussian random variables is treated. First the correlated variables are de-correlated using the Karhunen-Loève transform and then each of the transformed random variables is normalized by its variance so that they can be quantized using Max-Lloyd quantizers [118, 109] designed for unit variance Gaussian random variables. Quantizers with different numbers of reconstruction levels are available for each normalized random variable and hence each random variable with its associated set of quantizers can be characterized by its QF. The bit allocation problem is then to find a quantizer for each random variable so that the distortion (in this case, the mean squared error) is minimized subject to a bit rate budget.

The solution to the bit allocation problem is based on a simple exponential model of the QF, or in other words, a continuous function is used to model the relationship between the bit rate and the resulting distortion. Therefore the bit allocation problem can be formulated as a continuous constrained optimization problem and the solution can be found using the Lagrangian multiplier method [50], since the derivatives can be found explicitly. In this approach, the resulting bit assignment is not constrained to positive integer values but since the quantizers used operate only for positive integers, some trial and error procedure is used to adjust the bit rate assignment.

In [187] a similar problem is studied and again a continuous model for the QF is used. The main improvement of this algorithm is that the QF does not need to be exponential. The assumptions necessary for the proposed bit allocation are that the model QF is strictly convex and decreasing. Also the bit assignment is constrained to be non-negative. Clearly the constraints on the model QF are much weaker in this approach than in the one presented previously and hence a better modeling of the real QF is possible.

The above two approaches to optimal bit allocation are typical examples of a model-based bit allocation scheme. In both cases the QF, which is a set of rate distortion points, is modeled using a continuous function. This is more in the spirit of RDT than ORDT, since an elegant approximation to the discrete reality is used to guide the search for a solution. The advantage of this approach is that continuous optimization theory can be employed to find the solution. On the other hand, this approach has the disadvantages that the bits allocated to the particular quantizers are real numbers and that the allocation is only as good as the model used for the QF. In Addition, both approaches suffer from the requirement that all quantizers employed must have the same QF.

In [187] it is noted that if the bit assignment needs to be integer, then the approach presented in [54] could be used. In [54] a general algorithm for discrete resource allocation is presented, which is based on marginal analysis of incremental return. This algorithm has several nice features. First it does not require that the QFs of different sources need to be identical. Second it will result in bit allocations which are positive integers. Third, it achieves the allocation of M bits in M steps.

The main disadvantage of this algorithm is that a model of the QF is still necessary, which has to be defined for all positive integers values. This model QF needs to be convex, where convexity is defined as having decreasing first differences, and the model QF also needs to be strictly decreasing.

In [208] an extension to the approach presented in [54] is proposed, which attempts to drop the requirement for convex model QFs. Nevertheless model QFs are still required since they need to be evaluated at all positive integer values. Furthermore, the model QFs need to be non-increasing functions. The algorithm invokes first the scheme presented in [54], which is known to be optimal in the case of convex QFs. Since no assumption about the convexity of the QFs is made, the resulting solution is tested for optimality by employing a quasi-convexity condition. It is shown in [208] that if this condition is satisfied, the solution is still optimal even though the QFs are not convex. When the quasi-convexity condition is violated, the author suggests to adjust the solution by a paired-coordinate correction scheme, in the hope that a solution can be found which does satisfy the quasi-convexity condition.

If this correction fails, another procedure can be invoked which was original proposed in [207], which is based on dynamic programming. The bit allocation algorithm proposed in [207] is strongly related to the well known "knapsack" problem [132]. In the "knapsack" problem, a robber has to fill a bag of a given size with goods of different sizes and different values. The goal is to fill the bag in such a fashion that the final value is maximized. In the proposed optimal bit allocation algorithm, the available bits take the role of the bag size and the quantizers take the role of the goods, where the sizes of the goods is related to the rates required by the quantizers and the values of the goods relates to the resulting distortions.

In [190] an algorithm for optimal bit allocation is proposed which only assumes that the QFs are defined for all integers in a given range, even though the underling theory does not require this restriction. This algorithm is based on the Lagrangian multiplier method which we will discuss in some detail in the next section. The great advantage of the algorithm is that no assumptions about the QFs are made and hence an "ultimate bit allocation" (UBA) is possible. The meaning of UBA is that for every outcome of the source, all admissible quantizers are employed and the QF is recorded on-line. This has not been possible with the schemes presented earlier, since model QFs where always necessary, which usually are obtained by statistical means. Even though the underlying theory of the Lagrangian multiplier method does not require the QFs to contain only integer rates, nor that the QFs can be evaluated for all integer rates between a lower and an upper bound, these are the assumption made in [190].

In [222] the first part of the algorithm in [208], which is based on the algorithm in [54] is extended to enable a bit assignment of positive real numbers. The argument for this extension from the positive integers to the positive real numbers is based on the fact that when entropy coding of the quantizer output is employed, the average bit rate will be a real number. The main novelty of this approach is that no QF models are required, since it is not assumed that the QF can be evaluated for all integer values.

The goal of the algorithm in [222] is to find quantizers which belong to the convex hull of the QF. Since the Lagrangian multiplier method also finds the quantizers on the QF, it is clear that this algorithm and the one proposed in [190] are very similar. By restricting the solutions to be on the convex hull, neither the paired-coordinate correction scheme nor the dynamic programming scheme proposed in [208] are necessary. On the other hand the obtained solution might be slightly sub-optimal.

In [191] the earlier work in [190] is extended. It is explicitly stated that the restriction to the positive integers is not fundamental to the approach and that the Lagrangian multiplier method also works for any finite set of positive real numbers. However, the development of the algorithm uses this restriction.

In [29] an optimal pruning procedure for tree-structured source coding is introduced. The scheme is based on an algorithm called BFOS (after its inventors, Breiman, Friedman, Olshen and Stone), which was developed for the optimal pruning in tree-structured classification and regression [21]. It is shown in [29] that this scheme can be used to find tree-structured quantizers which are optimal in the rate distortion sense. It can also be used to find an optimal quad-tree decomposition which is a topic we briefly discussed in chapter 2 and will pick up again in chapter 4.

If the tree is defined in a special way, the BFOS algorithm lends itself to optimal bit allocation. This was first proposed in [29] and in [169] the algorithm proposed in [222] is restated using the BFOS algorithm. In [169] it is shown that when the QFs are convex, the BFOS results in a simple and low complexity procedure to find the optimal bit allocation. It is interesting to notice that the BFOS algorithm is also based on the Lagrangian multiplier method and hence the quantizers selected by BFOS for non-convex QFs are all on the convex hull. In the case of tree structured vector quantizers which are intended for low bit rate use, the convex hull solutions might not be good enough. To solve this problem, in [91] a modification is proposed which results also in quantizer selections which are above the convex hull.

So far the schemes presented dealt with QFs which are independent of each other. In other words, the selection of a specific quantizer for a given source does not change the QF of any other source. In practice, there are many systems where this is the case; on the other hand, there are also schemes where this is not the case. A good example for such a scheme is the encoding of a video sequence using a motion compensated video coder as discussed in chapter 2. The QF of the current frame strongly depends on the quantizer used for the previous frame since the previously reconstructed frame is used to predict the current one.

In [162] the above discussed concept of dependent quantization is introduced and extended in [163]. A solution to the problem of optimal bit allocation among depended quantizers is formulated which is based on the Lagrangian multiplier approach introduced in [191]. Unfortunately, since, for the general formulation, complete forward dependency is assumed, i.e., all the future QFs depend on the currently selected quantizer, the unconstrained problem has to be solved by an exhaustive search. In the case where an exhaustive search is used for the solution of the unconstrained problem, the Lagrangian multiplier method is not really appropriate, since it can only find the convex hull approximation, whereas a direct exhaustive search of the constrained problem would result in the optimal solution, having the same time complexity as the exhaustive search needed for the unconstrained problem. In fact, since the unconstrained problem has to be solved iteratively, the time complexity of the Lagrangian approach is even higher than that of the exhaustive search.

The main contribution of [162] are the specific dependent quantization frameworks studied, i.e., the optimal bit allocation among the frames of an MPEG sequence and optimal bit allocation among the levels of a multi-resolution coder. In the MPEG case, the I frames decouple the temporal dependencies and the Viterbi algorithm can be used to solve the unconstrained problem. The Viterbi algorithm is a dynamic programming (DP) scheme and since we use DP extensively in the later part of this monograph, we present a brief introduction to DP in section 3.4.

In addition to the Viterbi algorithm, which solve the unconstrained problem efficiently for a first order dependency, a pruning scheme is suggested to reduce the number of possible paths. The pruning scheme is based on an observed monotonicity hypothesis, which stipulates that the finer the quantization of the previous frame is, the more efficient the encoding of the prediction error becomes in the rate distortion sense.

In the next section we will discuss the general concept behind the Lagrangian multiplier approach and we will point out that conceptually any method can be used to solve the relaxed problem, not only dynamic programming. When dynamic programming is used, any order of dependency can be treated using a higher order state space.

Since the publication of [163], several other contributions are completely or partly based on the concept of dependent quantization. [97, 165, 143, 142, 224, 225]. Part of the material presented in this monograph also employs the Lagrangian multiplier method and dynamic programming, although we take a discrete optimization point of view on the problem of optimal bit allocation and show how this combination follows very naturally. This point of view also enables us to formulate the dependent quantizations scheme for higher order dependencies and to introduce a new scheme which solves the minimum maximum distortion problem for dependent quantization.

The minimum maximum bit allocation problem, is the problem of selecting the quantizers in such a way that the maximum distortion is minimized. It is studied in [105] and a solution which is based on an iterative decent procedure is presented for independent QFs. The procedure is quite simple in that one starts by giving zero bits to each source and then allocates enough bits to the source with the highest distortion so that its distortion is reduced by the smallest amount possible, using the QF of that source. Then the source with the largest distortion is found and again bits are allocated to that source until its distortion drops. This is repeated until all bits are used up. In [105] an extension of this scheme to dependent sources is outlined using a dual formulation described in [110], but it is deemed as formidably complicated due to the number of variables involved. In chapter 4 we propose a minimum maximum distortion optimal bit allocation algorithm for dependent quantizers which is not based on the dual formulation, but uses an iterative scheme which invokes a dynamic programming algorithm. We show that the proposed algorithm converges to the optimal solution.

It is interesting to notice that in [105] a solution to the minimum total distortion problem for dependent quantization is also outlined. The proposed solution also uses the Lagrangian multiplier method and the Viterbi algorithm and is therefore quite similar to the one proposed in [162, 163].

3.3 LAGRANGIAN MULTIPLIER METHOD

In this section we review the basic ideas behind the Lagrangian multiplier method, which is a mathematical tool to solve constrained optimization problems. The Lagrangian multiplier method [50] is well known for solving constrained minimization problems in a continuous framework. In this introduction we present the theory developed in [45], which shows that the Lagrangian multiplier method is also very valuable for constrained discrete optimization problems. Note that the Lagrangian multiplier method for resource allocation, which is discussed in this introduction, is closely related to the Lagrangian relaxation method [58, 48, 49] for integer programming. This is in fact where we first encountered the Lagrangian relaxation idea. It basically means, that one adds the constraint which makes the problem hard to solve to the objective function, weighted by the Lagrangian multiplier. Then the new problem, called the relaxed problem, is solved iteratively to find the Lagrangian multiplier which results in the tightest bound, which might also be the optimal solution. In ORDT the main aim is to optimally select the available quantizers so that the resulting distortion is minimized for the available bit rate. This is clearly a constrained optimization problem, and since there are only a finite number of decisions which can be made by the encoder, the integer version of the Lagrangian multiplier method is used throughout this monograph. In the following development of the Lagrangian multiplier method, we follow the original paper [45], but put the general theory into the framework of ORDT.

3.3.1 Main theorem

Theorem 1 *Let S_B be a finite set and $B \in S_B$ be a member of that set. Let $R(B)$ and $D(B)$ be real valued functions defined over S_B. Then, for any $\lambda \geq 0$ the optimal solution $B^*(\lambda)$ to the unconstrained problem,*

$$\min_{B \in S_B} (D(B) + \lambda * R(B)), \tag{3.15}$$

is also an optimal solution to the constrained problem,

$$\min_{B \in S_B} D(B), \quad \textit{subject to:} \quad R(B) \leq R(B^*(\lambda)). \tag{3.16}$$

Proof (by contradiction): Assume that the above theorem is false. This implies that there exists a $B \in S_B$ such that $D(B) < D(B^*(\lambda))$ and $R(B) \leq R(B^*(\lambda))$. Hence $D(B) + \lambda \cdot R(B) < D(B^*(\lambda)) + \lambda \cdot R(B^*(\lambda))$, which is a contradiction since $B^*(\lambda)$ is the optimal solution to the unconstrained problem. ■

The theorem does not guarantee a solution to the constrained problem. All it says is that to every non-negative λ, there exists a corresponding constrained problem whose solution is identical to that of the unconstrained problem. The main point is, however, that if $R(B^*(\lambda))$ happens to be equal to an upper bound R_{max}, $B^*(\lambda)$ is the desired solution to the constrained problem,

$$\min_{B \in S_B} D(B), \quad \text{subject to:} \quad R(B) \leq R_{max}. \tag{3.17}$$

A problem which remains to be solved is how to find a λ such that $R(B^*(\lambda)) = R_{max}$. We will address that problem in section 3.3.3. When $D(B)$ is the distortion for encoding a given source using a certain quantizer B and $R(B)$ the corresponding rate, then Eq. (3.17) represents the mathematical formulation of the optimal bit allocation problem where R_{max} is the number of available bits.

3.3.2 λ theorem

Theorem 2 *If $R(B^*(\lambda_1)) > R(B^*(\lambda_2))$ then,*

$$\lambda_2 \geq -\frac{D(B^*(\lambda_1)) - D(B^*(\lambda_2))}{R(B^*(\lambda_1)) - R(B^*(\lambda_2))} \geq \lambda_1. \tag{3.18}$$

Proof: By the optimality of $B^*(\lambda_1)$ the following holds,

$$D(B^*(\lambda_1)) + \lambda_1 \cdot R(B^*(\lambda_1)) \leq D(B^*(\lambda_2)) + \lambda_1 \cdot R(B^*(\lambda_2)). \tag{3.19}$$

Solving for λ_1 and keeping in mind that $R(B^*(\lambda_1)) > R(B^*(\lambda_2))$ results in,

$$-\frac{D(B^*(\lambda_1)) - D(B^*(\lambda_2))}{R(B^*(\lambda_1)) - R(B^*(\lambda_2))} \geq \lambda_1, \tag{3.20}$$

which proves the right hand side of the theorem. The left hand side can be proven in a similar fashion. By the optimality of $B^*(\lambda_2)$ the following holds,

$$D(B^*(\lambda_2)) + \lambda_2 \cdot R(B^*(\lambda_2)) \leq D(B^*(\lambda_1)) + \lambda_2 \cdot R(B^*(\lambda_1)). \tag{3.21}$$

Solving for λ_2 and keeping in mind that $R(B^*(\lambda_1)) > R(B^*(\lambda_2))$ results in,

$$\lambda_2 \geq -\frac{D(B^*(\lambda_1)) - D(B^*(\lambda_2))}{R(B^*(\lambda_1)) - R(B^*(\lambda_2))}. \tag{3.22}$$

■

The λ theorem states that given two optimal solutions, the ratio of the change in optimum distortion to the change in the required rate is bounded between the two multipliers. In particular, if the set of solutions produced by Lagrangian multipliers results in a distortion that is a differentiable function of the rate at some point, then it follows from the above theorem that λ at that point is the derivative of the distortion with respect to the rate,

$$-\frac{dD}{dR} = \lambda. \tag{3.23}$$

By applying this result to the ORDF (assuming the derivative exists), the following holds,

$$\frac{dR}{dD} = \frac{-1}{\lambda}. \tag{3.24}$$

The λ theorem is the basis of a fast convex search for the optimal λ introduced in [164], whereas Eq. (3.24) and the knowledge that a continuous rate distortion function is convex is the basis for the very fast convex search we propose in chapter 4.

3.3.3 Iterative Search for λ

Theorem 3 *$R(B^*(\lambda))$ is a non-increasing function of the Lagrangian multiplier λ.*

Proof (by contradiction): Let $\lambda_2 > \lambda_1 > 0$ and assume that $R(B^*(\lambda_2)) > R(B^*(\lambda_1))$. From the λ theorem it follows that $\lambda_1 \geq \lambda_2$, which is a contradiction. Therefore we conclude that $R(B^*(\lambda_2)) \leq R(B^*(\lambda_1))$. ■

Corollary 1 *$D(B^*(\lambda))$ is a non-decreasing function of the Lagrangian multiplier λ.*

Proof: From Eq. (3.19) it follows that $R(B^*(\lambda_1)) \geq R(B^*(\lambda_2))$ implies that $D(B^*(\lambda_1)) \leq D(B^*(\lambda_2))$. Eq. (3.21) indicates that $D(B^*(\lambda_1)) \leq D(B^*(\lambda_2))$ implies $R(B^*(\lambda_1)) \geq R(B^*(\lambda_2))$. Therefore $R(B^*(\lambda_1)) \geq R(B^*(\lambda_2))$ if and only if $D(B^*(\lambda_1)) \leq D(B^*(\lambda_2))$. ■

Having established that $R(B^*(\lambda))$ is a non-increasing function of λ, the Lagrangian multiplier, the bisection method can be used to find the optimal λ (see Fig. 3.3). The bisection method starts with two initial guesses for λ, λ_l and λ_u which have to be selected such that $R(B^*(\lambda_l)) \geq R_{max} \geq R(B^*(\lambda_u))$, where R_{max} is the maximum number of bits available, i.e., the target rate.

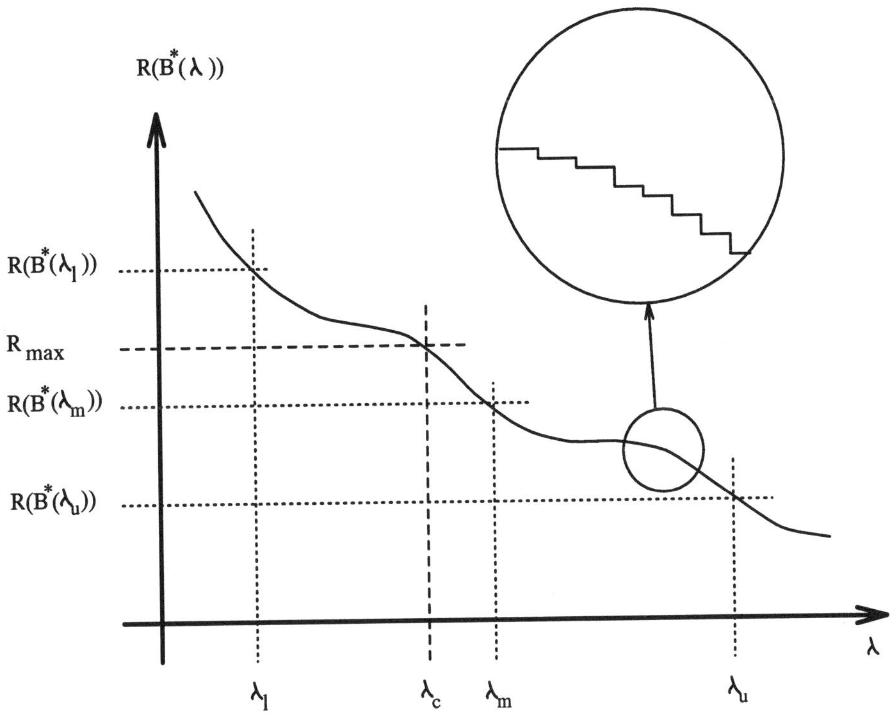

Figure 3.3 Bisection method applied to non-increasing $R(B^*(\lambda))$

Because of the non-increasing property of $R(B^*(\lambda))$, the desired λ (if it exits) lies between λ_l and λ_u. By selecting a $\lambda_m = \frac{\lambda_l + \lambda_u}{2}$ the initial interval is bisected. Then $R(B^*(\lambda_m))$ is evaluated and if $R(B^*(\lambda_m)) \geq R_{max}$ then $\lambda_l = \lambda_m$, otherwise $\lambda_u = \lambda_m$. By repeating this procedure, tighter and tighter bounds for the required λ are found. As mentioned before, there might not be a λ which satisfies $R(B^*(\lambda)) = R_{max}$ exactly, since R is defined on the finite set S_B. Therefore, the bisection is stopped when a solution is found which satisfies $R_{max} - \epsilon \leq R(B^*(\lambda)) \leq R_{max} + \epsilon$, where $\epsilon \geq 0$.

When the solution to the relaxed problem can be found independently for each QF, more elaborate search strategies can be used. In fact if one uses the Lagrangian multiplier method to analyze the optimal bit allocation schemes reviewed in section 3.2.1, the search algorithm to find the optimal λ is the main difference between the different approaches. In chapter 4 we present a very fast convex search to find the optimal λ. The proposed search is based on Eq. (3.24) and the knowledge that a continuous rate distortion function is convex.

3.3.4 Convex hull solutions

From RDT we know that the RDF is a non-increasing convex function. We have shown is section 3.2 that the ORDF is a set of points in the rate distortion space. As can be seen from Fig. 3.2, when the points of the ORDF are connected by straight lines, the resulting curve is not necessarily convex. In this section we show that the Lagrangian multiplier method can only be used to find the points of the ORDF which belong to the convex hull of the ORDF, which is also indicated in Fig. 3.2.

Assume that $B^*(\lambda)$ is an optimal solution to $min_{B \in S_B}(D(B) + \lambda \cdot R(B))$. This implies the following,

$$D(B^*(\lambda)) + \lambda \cdot R(B^*(\lambda)) \leq D(B) + \lambda \cdot R(B), \tag{3.25}$$

where $B \in S_B$ can be any admissible decision vector. By rearranging terms, this expression can be written as follows,

$$R(B) \geq -\frac{1}{\lambda} \cdot D(B) + \left\{ R(B^*(\lambda)) - \frac{1}{\lambda} \cdot D(B^*(\lambda)) \right\}. \tag{3.26}$$

Consider the case when the equal sign in Eq. (3.26) holds. Then Eq. (3.26) defines a line in the rate distortion plane (see Fig. 3.4). Eq. (3.26) states that all points (D(B),R(B)) of the QF must lie in the upper right half-plane of the rate distortion plane, i.e., above the line defined by Eq. (3.26), (see Fig. 3.4), whereas the set of optimal solutions $\{(D(B_i^*(\lambda)), R(B_i^*(\lambda)))\}$ is on the line. Recall that the slope of the line equals $\frac{-1}{\lambda}$, the rate $R(B^*(\lambda))$ is a non-increasing function of λ and the distortion $D(B^*(\lambda))$ is a non-decreasing function of λ. Therefore if we sweep λ from zero to infinity, and connect the points $(D(B^*(\lambda)), R(B^*(\lambda)))$ by straight lines, the convex hull of the ORDF is generated. A point of the ORDF which is not on the convex hull cannot be found using a Lagrangian approach since one cannot find a line (and hence a λ) which partitions the plane such that all solutions belong to the resulting upper right half-plane.

Note that in the entire discussion of the Lagrangian multiplier method we have not specified how the relaxed problem of Eq. (3.15) is solved. For the

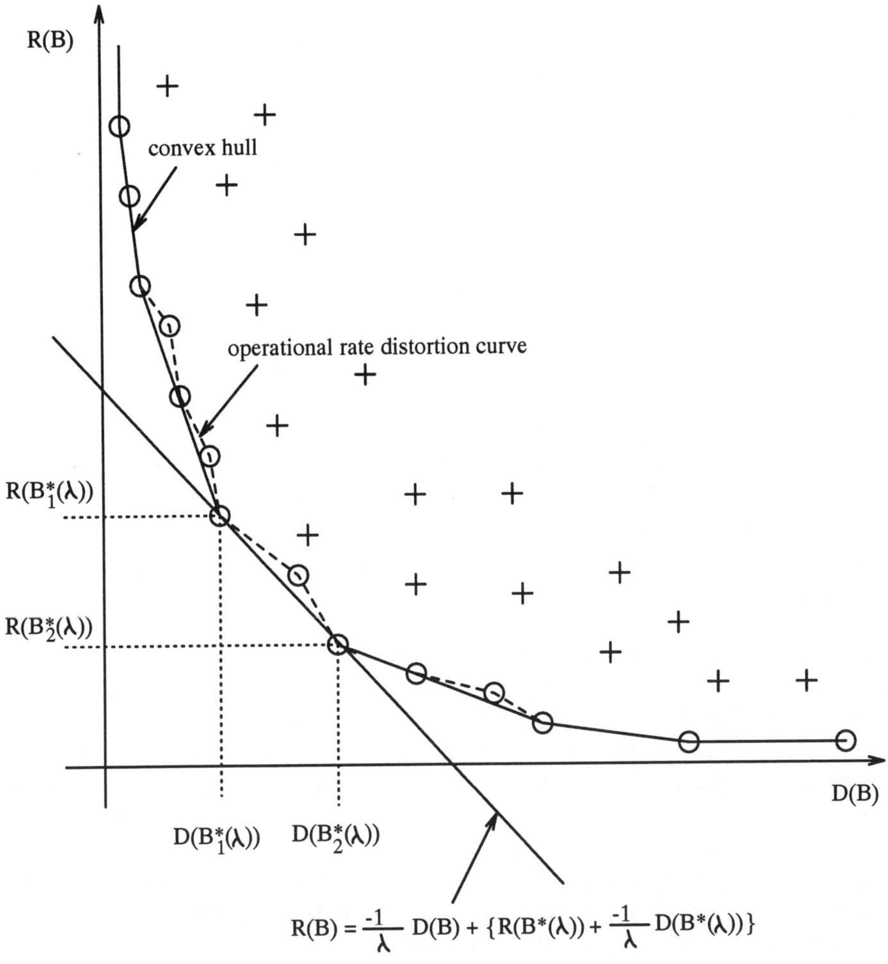

Figure 3.4 Rate distortion plane. The circles indicate the operational rate distortion function. Note that not all circles belong to the convex hull. These points are inaccessible for the Lagrangian multiplier method.

Lagrangian multiplier method to work, the only requirement is that the optimal solution of the relaxed problem can be found. This insight enables us to formulate an optimal bit allocation procedure for dependent quantizers with arbitrarily high dependencies in chapter 4.

It is important to notice that the roles of the rate and the distortion can be interchanged and the presented theory still applies. In other words, the following problem,

$$\min_{B \in S_B} R(B), \qquad \text{subject to:} \quad D(B) \leq D_{max}, \tag{3.27}$$

can also be solved by the Lagrangian multiplier method.

3.4 DYNAMIC PROGRAMMING

In this section, we give a short review of deterministic dynamic programming (DP). DP is a very general tool in the field of optimization and has been studied extensively [12, 14]. This short presentation of DP is restricted to the properties and algorithms used in the remainder of this monograph and it is by no means a comprehensive overview. In the field of Electrical Engineering, the most well known DP algorithm is the Viterbi decoder [53].

DP is a powerful tool for optimization problems which can be separated into a past, a present and a future, where given the present, the future is independent of the past. The knowledge represented in the present is usually referred to as the state. There are deterministic and stochastic DP schemes. A famous example for stochastic DP is a Kalman filter where the objective is to minimize the mean squared error. We will focus on deterministic DP and we will give a shortest path example later on. The main advantage of DP over other conventional optimization techniques is the fact that DP can handle difficult constraint sets such as integer or discrete sets. Furthermore DP leads to a globally optimal solution as opposed to variational techniques, for which this cannot be guaranteed in general. As will be shown later, both of these features are of importance for the schemes developed in the remainder of this monograph.

3.4.1 Problem formulation

In this monograph we are interested in applying DP to minimize objective functions of the following form,

$$G(v_0, \ldots, v_{N-1}) = \sum_{i=0}^{N-1} g_i(v_{i-a}, \ldots, v_{i+b}), \tag{3.28}$$

where a and b are non-negative integers, and the v_i represent the decision variables which belong to finite sets of admissible decision variables V_i, $v_i \in V_i$. The decision variables $v_{-a}, \ldots, v_{-1}$ and $v_N, \ldots, v_{N-1+b}$ are the boundary parameters of the sum and can be set to any desired value. Eq. (3.28) express the fact that the objective function $G(\cdot)$ is the sum of local objective functions $g_i(v_{i-a}, \ldots, v_{i+b})$. Each of the local objective functions depends only on a subset of size $a + b + 1$, $(v_{i-a}, \ldots, v_{i+b})$, of all decision variables arranged around the local decision variable v_i. We call this subset of decision variables the neighborhood i.

The goal is to find the set of decision variables which results in the smallest objective function,

$$\min_{[v_0,\ldots,v_{N-1}] \in V_0 \times \ldots \times V_{N-1}} G(v_0, \ldots, v_{N-1}). \tag{3.29}$$

As we will see in the following sections, because of the special structure of the objective function, DP can be used to reduce this N-dimensional optimization problem into N $(a + b + 1)$-dimensional optimization problems.

3.4.2 DP recursion formula

Since we deal with a finite number of possible combinations of decision variables, the optimization problem of Eq. (3.29) can be solved by an exhaustive search. The time complexity for such an exhaustive search is $O(|V_i|^N)$, where $|V_i|$ denotes the cardinality of V_i, and we assume that all V_is have the same cardinality. We will show that the proposed algorithm reduces this complexity significantly. Note that when we use the term time complexity, we refer to the number of comparisons necessary to find the optimal solution.

We denote by $g_l^*(v_{l-a+1}, \ldots, v_{l+b})$ the minimum cost up to and including neighborhood l, that is

$$g_l^*(v_{l-a+1}, \ldots, v_{l+b}) = \min_{[v_0,\ldots,v_{l-a}] \in V_0 \times \ldots \times V_{l-a}} \sum_{i=0}^{l} g_i(v_{i-a}, \ldots, v_{i+b}). \tag{3.30}$$

From Eq. (3.30) it follows that,

$$
\begin{aligned}
& g_{l+1}^*(v_{l+1-a+1}, \ldots, v_{l+1+b}) \\
= & \min_{[v_0,\ldots,v_{l+1-a}] \in V_0 \times \ldots \times V_{l+1-a}} \sum_{i=0}^{l+1} g_i(v_{i-a}, \ldots, v_{i+b}) \qquad (3.31) \\
= & \min_{v_{l+1-a} \in V_{l+1-a}} \Bigg[\min_{[v_0,\ldots,v_{l-a}] \in V_0 \times \ldots \times V_{l-a}} \\
& \left(\sum_{i=0}^{l} g_i(v_{i-a}, \ldots, v_{i+b}) + g_{l+1}(v_{l+1-a}, \ldots, v_{l+1+b}) \right) \Bigg]. \qquad (3.32)
\end{aligned}
$$

Since $g_{l+1}(v_{l+1-a}, \ldots, v_{l+1+b})$ does not depend on $v_0, \ldots, v_{l-a}$, it can be moved outside the inner minimization. Then the resulting inner minimization is equal to $g_l^*(v_{l-a+1}, \ldots, v_{l+b})$ in Eq. (3.30) and the following DP recursion formula results,

$$
\begin{aligned}
& g_{l+1}^*(v_{l+1-a+1}, \ldots, v_{l+1+b}) \\
= & \min_{v_{l+1-a} \in V_{l+1-a}} [g_l^*(v_{l+1-a}, \ldots, v_{l+b}) + g_{l+1}(v_{l+1-a}, \ldots, v_{l+1+b})] \qquad (3.33)
\end{aligned}
$$

This recursive formula requires only $|V_i|^{a+b+1}$ comparisons (we assume that all V_is have the same cardinality $|V_i|$) to find $g_{l+1}^*(v_{l+1-a+1}, \ldots, v_{l+1+b})$, $\forall$ $[v_{l+1-a+1}, \ldots, v_{l+1+b}] \in V_{l+1-a+1} \times \ldots \times V_{l+1+b}$, given $g_l^*(v_{l+1-a}, \ldots, v_{l+b})$, which is a substantial saving from the exhaustive search approach.

3.4.3 Forward DP algorithm

Having established the DP recursion formula, the forward DP algorithm can be used to find the optimal solution to problem (3.29). Note that the forward DP algorithm is also called the Viterbi algorithm [53]. First, the recursion needs to be initialized,

$$
\begin{aligned}
g_{a-1}^*(v_0, \ldots, v_{a+b-1}) = \sum_{i=0}^{a-1} g_i(v_{i-a}, \ldots, v_{i+b}) \\
\forall\, [v_0, \ldots, v_{a+b-1}] \in V_0 \times \ldots \times V_{a+b-1}. \qquad (3.34)
\end{aligned}
$$

We introduce a back pointer which will be used to remember the optimal selection,

$$
i_{a-1}(v_0, \ldots, v_{a+b-1}) = [v_{-a}, \ldots, v_{b-1}]. \qquad (3.35)
$$

Next, the recursion is started, hence the DP recursion formula (3.32) is applied for $l = a - 1$ up to and including $l = N - 2$, that is,

$$\begin{aligned} & g^*_{l+1}(v_{l+1-a+1}, \ldots, v_{l+1+b}) \\ & = \min_{v_{l+1-a} \in V_{l+1-a}} \left[g^*_l(v_{l+1-a}, \ldots, v_{l+b}) + g_{l+1}(v_{l+1-a}, \ldots, v_{l+1+b})\right], \\ & \quad \forall \left[v_{l+1-a+1}, \ldots, v_{l+1+b}\right] \in V_{l+1-a+1} \times \ldots \times V_{l+1+b}. \end{aligned} \tag{3.36}$$

Again the back pointer is assigned, for which the argument which minimizes the above problem needs to be known,

$$\begin{aligned} & v^*_{l+1-a}(v_{l+1-a+1}, \ldots, v_{l+1+b}) \\ & = \arg \min_{v_{l+1-a} \in V_{l+1-a}} \left[g^*_l(v_{l+1-a}, \ldots, v_{l+b}) + g_{l+1}(v_{l+1-a}, \ldots, v_{l+1+b})\right], \\ & \quad \forall \left[v_{l+1-a+1}, \ldots, v_{l+1+b}\right] \in V_{l+1-a+1} \times \ldots \times V_{l+1+b}, \end{aligned} \tag{3.37}$$

therefore,

$$\begin{aligned} & i_{l+1}(v_{l+1-a+1}, \ldots, v_{l+1+b}) \\ & = \left[v^*_{l+1-a}(v_{l+1-a+1}, \ldots, v_{l+1+b}), v_{l+1-a+1}, \ldots, v_{l+b}\right], \\ & \quad \forall \left[v_{l+1-a+1}, \ldots, v_{l+1+b}\right] \in V_{l+1-a+1} \times \ldots \times V_{l+1+b}. \end{aligned} \tag{3.38}$$

Then the final solution is found by observing that,

$$\begin{aligned} & \min_{[v_0, \ldots, v_{N-1}] \in V_0 \times \ldots \times V_{N-1}} G(v_0, \ldots, v_{N-1}) \\ & = \min_{[v_{N-a}, \ldots, v_{N-1}] \in V_{N-a} \times \ldots \times V_{N-1}} g^*_{N-1}(v_{N-a}, \ldots, v_{N-1+b}). \end{aligned} \tag{3.39}$$

The final back pointer requires the arguments which minimize Eq. (3.38),

$$[v^*_{N-a}, \ldots, v^*_{N-1}] = \arg \min_{[v_{N-a}, \ldots, v_{N-1}] \in V_{N-a} \times \ldots \times V_{N-1}} g^*_{N-1}(v_{N-a}, \ldots, v_{N-1+b}). \tag{3.40}$$

Using these arguments, the following final back pointer results,

$$i_N = [v^*_{N-a}, \ldots, v^*_{N-1}, v_N, \ldots, v_{N-1+b}]. \tag{3.41}$$

Note that $v_N, \ldots, v_{N-1+b}$ are fixed parameters and hence they are identical to their optimal values $v^*_N, \ldots, v^*_{N-1+b}$. Using the described approach, we can find the minimum value of the objective function, but we also need to know the optimal decision variables $v^*_0, \ldots, v^*_{N-1}$. These decision variables can be found by following the back pointers during the backtracking stage. Note that i_N identifies the optimal decision variables $v^*_{N-a}, \ldots, v^*_{N-1+b}$. The remaining decision variables can be found as follows. For $l = N - 1, \ldots, a$

$$v^*_{l-a} = \left[i_l(v^*_{l-a+1}, \ldots, v^*_{l+b})\right]_1, \tag{3.42}$$

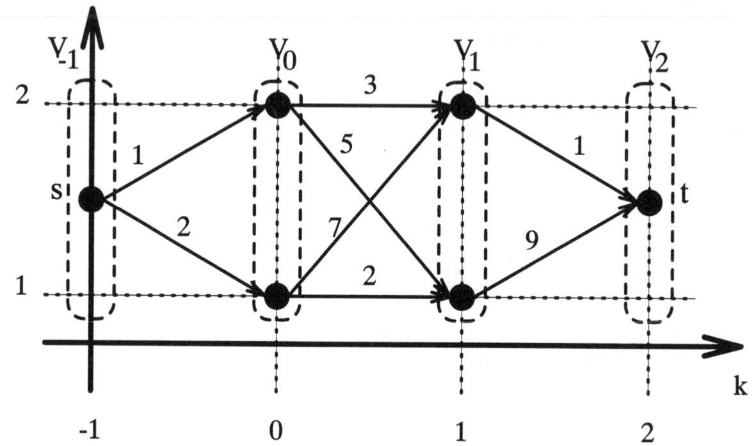

Figure 3.5 Example trellis.

where $[\cdot]_1$ refers to the first element in the vector.

The time complexity for the exhaustive search is exponential. The time complexity for the DP approach depends directly on the size of the neighborhood and is $O(N * |V_i|^{a+b+1})$, where we again assume that all V_is have the same cardinality.

3.4.4 Example

We now consider a simple example to illustrate the above points. In this example we make the neighborhood as small as possible by setting a=1, and b=0. For this special configuration it is easy to draw a trellis (see Fig. 3.5) to illustrate the algorithm. The trellis of Fig. 3.5 completely specifies the optimization problem. The dashed boxes indicate the admissible decision vector sets, $V_1 = \{1, 2\}$ and $V_2 = \{1, 2\}$, and the black circles represent a particular member of these sets. The start node s at epoch -1 and the terminal node t at epoch 2 are the boundary parameters of the sum, as discussed earlier. The numbers on top of the arrows indicate the local objective functions $g_i(v_{i-1}, v_i)$. For example $g_1(v_0 = 1, v_1 = 2) = 7$. The goal is to solve the following minimization problem,

$$\min_{[v_0,v_1]\in V_0\times V_1} G(v_0,v_1) = \min_{[v_0,v_1]\in V_0\times V_1} \sum_{i=0}^{2} g_i(v_{i-1},v_i), \tag{3.43}$$

where $v_{-1} = s$ and $v_2 = t$. Note that this minimization procedure is identical to finding the shortest path through the trellis.

We can now use Eq. (3.33) to initialize the forward DP, i.e.,

$$g_0^*(v_0) = g_0(v_{-1},v_0), \qquad \forall\, v_0 \in V_0, \tag{3.44}$$

which results in $g_0^*(1) = 2$ and $g_0^*(2) = 1$. The back pointers $i_0(1) = s$ and $i_0(2) = s$ are set according to Eq. (3.35). Next, the recursion is started, hence the DP recursion formula (3.32) is applied. First for $l = 0$,

$$g_1^*(v_1) = \min_{v_0\in V_0} \left[g_0^*(v_0) + g_1(v_0,v_1)\right], \qquad \forall\, v_1 \in V_1, \tag{3.45}$$

which results in $g_1^*(1) = 4$ and $g_1^*(2) = 4$ and the back pointers are $i_1(1) = 1$ and $i_1(2) = 2$. Now the DP recursion formula is applied for $l = 1$,

$$g_2^*(v_2) = \min_{v_1\in V_1} \left[g_1^*(v_1) + g_2(v_1,v_2)\right], \qquad \forall\, v_2 \in V_2, \tag{3.46}$$

which results in $g_2^*(t) = 5$ and the back pointer is $i_2(t) = 2$.

The final solution can now be found using Eq. (3.38),

$$\min_{[v_0,v_1]\in V_0\times V_2} G(v_1,v_2) = \min_{v_2\in V_2=\{t\}} g_2^*(v_2) = 5, \tag{3.47}$$

which was clear since the cardinality of V_2 is one and hence there are no other competitors for the optimal solution. The optimal decision variables can be found by backtracking, $v_2^* = t$, $v_1^* = i_2(v_2^*) = 2$, $v_0^* = i_1(v_1^*) = 2$ and $v_{-1}^* = i_0(v_0^*) = s$.

3.5 SHORTEST PATH ALGORITHM

In this section we present an algorithm for finding a shortest path through a weighted directed acyclic graph which is very similar to the DP approach discussed in section 3.4. We will employ this algorithm extensively for the optimal encoding of object boundaries to be addressed in chapter 8.

A graph G=(V,E) is defined by its vertex set V and its edge set E. An edge $(u, v) \in E$ connects vertex $u \in V$ with vertex $v \in V$. In the case of a weighted graph, the edge (u, v) has a weight assigned, which is usually accomplished by a weight function $w(u, v) : E \to \mathcal{R}$. For directed graphs, the edge (u, v) and the edge (v, u) carry different weights, and quite often one of these weights is zero. A path of order K from vertex u to a vertex u' is an ordered set $\{v_0, \ldots, v_K\}$ such that $u = v_0$, $u' = v_K$ and $(v_{k-1}, v_k) \in E$ for $k = 1, \ldots, K$. The order of the path is the number of edges in the path. The length of a path is defined as follows,

$$\sum_{k=1}^{K} w(v_{k-1}, v_k). \tag{3.48}$$

In a directed graph, a path $\{v_0, \ldots, v_K\}$ forms a cycle if $v_0 = v_K$ and the path contains at least one edge. A directed path without cycles is called a directed acyclic graph or DAG for short.

An interesting weighted DAG is the trellis in Fig. 3.5, which was used to illustrate the forward DP algorithm for a neighborhood of size one (a=1, b=0). By considering the trellis a weighted DAG, the minimization problem stated in (3.43) can be considered a shortest path problem in a weighted DAG. There exists an efficient algorithm to find the shortest path through a weighted DAG. Following the notation of [34], we call this algorithm the DAG-shortest-path algorithm.

The first step in the DAG-shortest-path algorithm is to topologically sort the DAG. A topological sort of a DAG G=(V,E) is a linear ordering of all its vertices such that if G contains an edge (u, v), then u appears before v in the ordering. A topological sort of a DAG can be viewed as an ordering of its vertices along a horizontal line so that all directed edges go from left to right. It can be shown [34] that a DAG can be topologically sorted in $\Theta(V + E)$. Whenever we use a DAG, it is already in a topologically sorted order and hence we will not discuss the topological sort. In Fig. 3.6 the trellis from Fig. 3.5 has been topologically sorted. Note that we also numbered the vertices from zero to $N - 1$, where $N = 6$ in this example. This enumeration of the DAG simplifies the notation needed for the algorithm.

Let $g^*(u)$ represent the minimum cost to reach vertex u from the source vertex 0. Clearly $g^*(N - 1 = 5)$ is the solution to problem (3.43). Let $q(u)$ be a back pointer which is used to remember the optimal path. Then the DAG shortest path algorithm works as follows:

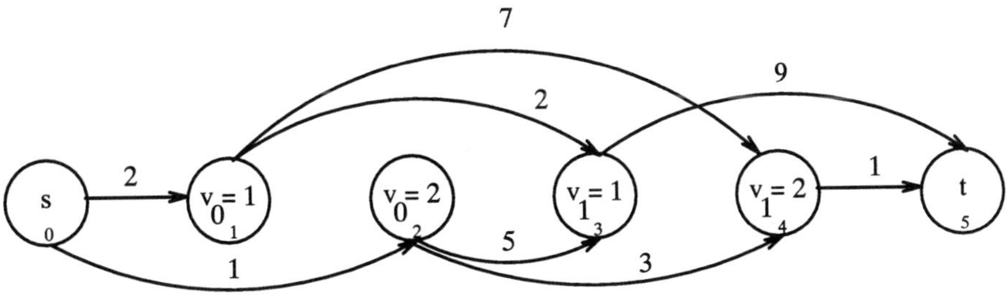

Figure 3.6 Topologically sorted trellis.

```
g*(0) = 0;
for i = 1,...,N − 1;
{
    g*(i) = ∞;
}
for i = 0,...,N − 2;
{
    for j = i + 1,...,N − 1;
    {
        if (g*(i) + w(i,j) < g*(j));
        {
            g*(j) = g*(i) + w(i,j);
            q(j) = i;
        }
    }
}
```

The optimal path $\{p_0^*,\cdots,p_{N-1}^*\}$ can be found by back tracking the pointers $q(i)$ in the following recursive fashion (by definition $p_{N-1}^* = t$ and $p_0^* = s$),

$$p_{k-1}^* = q(p_k^*), \qquad k = N-1,\ldots,2. \tag{3.49}$$

The formal proof of the correctness of the DAG-shortest-path algorithm can be found in [34]. We will reason intuitively how this approach works using the example shown in Fig. 3.7. In Fig. 3.7, the optimal cost $g^*(j)$ is displayed inside each vertex. It is the optimal cost to go form vertex 0 to vertex j, having established the optimal path between vertex 0 and vertex i. The back pointers $q(j)$ are indicated as dashed arrows.

In Fig. 3.7, the DAG (init) shows the DAG after initialization, which corresponds to line (1) through (5) in the above algorithm. The "for loop" in line (6) selects the "from" vertices i in topological order and the "for loop" in line (8) selects all possible "to" vertices j in topological order. Each DAG in Fig. 3.7 represents the situation for a given vertex i, after the inner loop over the j vertices has been completed. Since the i vertices are selected in topological order, the optimal path from 0 to i has been established when vertex i is selected. During the j loop, the best path from 0 to j which goes over i is established, using the fact that the path form 0 to i is optimal.

Hence these two loops select each edge in the edge set E exactly once. Therefore lines (9) to (15) are processed for every edge. The most important part of this algorithm is the comparison in line (10). Here we test if the new smallest cost, $g^*(i) + w(i,j)$, to reach vertex j, given that i was the previous vertex, is smaller than the smallest cost used so far to reach vertex j, $g^*(j)$. If this cost is indeed smaller, then it is assigned as the new smallest cost to reach vertex j, $g^*(j) = g^*(i) + w(i,j)$. We also assign the back pointer of vertex j, $q(j)$ to point to vertex i since this is the previous vertex used to achieve $g^*(j)$.

The final result of this algorithm can be seen in at the bottom of Fig. 3.7. Clearly the optimal cost to reach vertex 5 from vertex 0 is 5 and following the back pointers, the optimal decision variables are $0 \Rightarrow s$, $2 \Rightarrow v_0 = 2$, $4 \Rightarrow v_1 = 2$, $5 \Rightarrow t$. Note that this is the same result we obtained using DP.

The optimality of this algorithm is based on Bellman's optimality principle [12], which is also the foundation of DP. The optimality principle states that if the optimal path from a to c includes b then the sub-path from a to b has to be an optimal path to go from a to b and the sub-path from b to c has also to be an optimal path to go from b to c. The idea behind the above algorithm is to establish optimal sub-paths from 0 to i in sequence. First, the sub-path from 0 to 0 is optimal by definition, then the optimal sub-path from 0 to 1 is established, which then is used to find the optimal sub-path from 0 to 2, and so on.

3.6 SUMMARY

In this chapter we first gave an introduction to rate distortion theory (RDT). We then presented operational rate distortion theory (ORDT) and reviewed the optimal bit allocation problem. Finally the Lagrangian multiplier method, dynamic programming and the related shortest path algorithm for a weighted acyclic directed graph were presented. This represents the background material which will be extensively used in the remainder of the monograph.

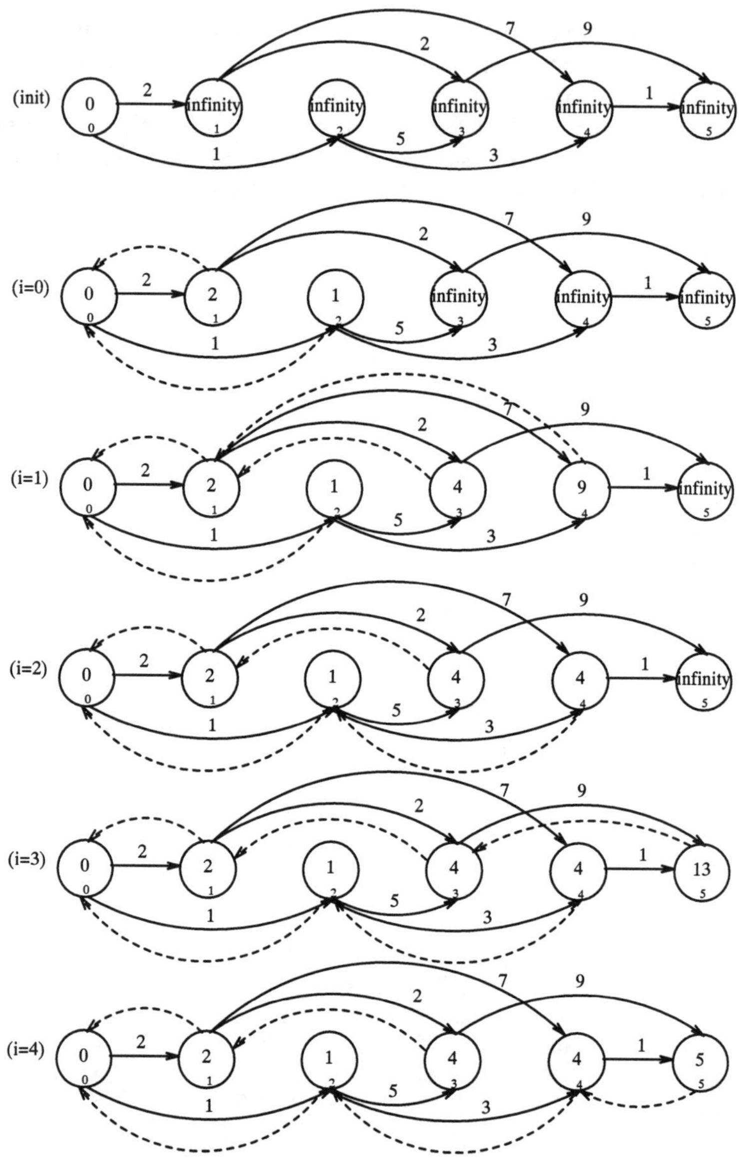

Figure 3.7 Different stages of the shortest path algorithm.

4

GENERAL CONTRIBUTIONS

In this chapter we present the general contributions of this monograph. These schemes and algorithms have been developed to solve specific problems, which are presented later on, but they are general enough that they have a much wider applicability than the problem they were originally developed for.

This chapter is organized as follows: In section 4.1 we formulate the solution for the optimal bit allocation problem for dependent quantizers, for arbitrary order of dependencies, when the objective is to minimize the total distortion. In section 4.2 we propose a very fast convex search to find the optimal Lagrangian multiplier λ. In section 4.3 we formulate the solution for the optimal bit allocation problem for dependent quantizers, for arbitrary order of dependencies, when the objective is to minimize the maximum distortion. In section 4.4 we derive a procedure and a sufficient and necessary condition for that procedure to find an optimal scanning path for an arbitrary quad-tree (QT) decomposition. In section 4.5 we propose an algorithm which finds the optimal QT decomposition in the rate distortion sense for a QT with leaf dependencies and in section 4.6 we summarize this chapter.

4.1 OPTIMAL BIT ALLOCATION FOR DEPENDENT QUANTIZERS USING THE MINIMUM TOTAL DISTORTION CRITERION

In this section, we formulate a general scheme to find the optimal bit allocation among dependent quantizers such that the total (average) distortion is minimized. We extend the approach for dependent quantizers introduced in [162, 163]. As we have pointed out in section 3.2.1, the general solution to the problem of optimal bit allocation among depended quantizers is formulated in [162, 163] assuming complete forward dependency, i.e., all future quantizer functions (QF) depend on the currently selected quantizer. The use of the Lagrangian multiplier method is proposed to solve the constrained problem by transforming it into a family of unconstrained problems. The idea behind Lagrangian relaxation is to transform the "hard" constrained problem into a a family of "easy" unconstrained problems. In the case of total dependency, even the unconstrained problems cannot be solved easily. In general, an exhaustive search is needed and hence the Lagrangian multiplier method is not really appropriate, since it can only find the convex hull approximation, whereas a direct exhaustive search of the constrained problem would result in the optimal solution.

In this section we assume that the current QF depends on a neighborhood of quantizers, which in the limiting cases can be all other quantizers or only the previous quantizer. Hence the proposed scheme contains the dependencies investigated in [162, 163] as special cases. We will show that he efficiency of the proposed approach directly depends on the size of the neighborhood.

4.1.1 Notation and assumptions

As stated above, we assume that the rate r_i and the distortion d_i of a source i depend on the quantizer selections of neighboring sources in a neighborhood defined by two non-negative integers a and b, that is,

$$R(x_0, \ldots, x_{N-1}) = \sum_{i=0}^{N-1} r_i(x_{i-a}, \ldots, x_{i+b}), \tag{4.1}$$

$$D(x_0, \ldots, x_{N-1}) = \sum_{i=0}^{N-1} d_i(x_{i-a}, \ldots, x_{i+b}), \tag{4.2}$$

where x_i is the quantizer of source i selected from the set of admissible quantizers X_i for source i and N is the total number of sources. As in section 3.4 the quantizers $x_{-a}, \ldots, x_{-1}$ and $x_N, \ldots, x_{N-1+b}$ are the boundary parameters of the sum and can be set to any desired value.

4.1.2 Problem formulation

The goal is to find the set of quantizers which results in the smallest total distortion for a given maximum bit rate R_{max}. Mathematically, this can be expressed as follows,

$$\min_{x_0,\ldots,x_{N-1}} D(x_0, \ldots, x_{N-1}) \quad \text{subject to:} \quad R(x_0, \ldots, x_{N-1}) \leq R_{max}. \tag{4.3}$$

We have shown in section 3.3 that the Lagrangian multiplier method can be used to solve constrained optimization problems of the above form.

4.1.3 Lagrangian multiplier approach

We define a Lagrangian source cost function, $j_{i,\lambda}(x_{i-a}, \ldots, x_{i+b})$, for every source,

$$j_{i,\lambda}(x_{i-a}, \ldots, x_{i+b}) = d_i(x_{i-a}, \ldots, x_{i+b}) + \lambda r_i(x_{i-a}, \ldots, x_{i+b}), \tag{4.4}$$

which then is used to define the total Lagrangian cost function,

$$J_\lambda(x_0, \ldots, x_{N-1}) = \sum_{i=0}^{N-1} j_{i,\lambda}(x_{i-a}, \ldots, x_{i+b}). \tag{4.5}$$

As we have pointed out in section 3.3, if we can find the global minimum of the total Lagrangian cost function of Eq. (4.5), then we can find the convex hull solution to the constrained problem of Eq. (4.3), using an iterative search, such as the one presented in section 4.2.

4.1.4 Solution of the unconstrained problem by dynamic programming

We have shown in section 3.4 that dynamic programming can be used to solve the following problem optimally and efficiently,

$$\min_{x_0,\ldots,x_{N-1}} G(x_0, \ldots, x_{N-1}), \tag{4.6}$$

when $G(x_0, \ldots, x_{N-1})$ has the structure given by Eq. (4.5) (see Eq. (3.28) and Eq. (3.29) in chapter 3). Hence dynamic programming (DP) can be used to solve the following unconstrained problem optimally,

$$\min_{x_0,\ldots,x_{N-1}} J_\lambda(x_0, \ldots, x_{N-1}). \tag{4.7}$$

We have also shown that the number of comparisons needed to find the optimal solution is $O(|X_i|^{a+b+1})$, where we assume that the cardinality of the set of admissible quantizers is the same for each source. Clearly in the important case where only the previous quantizer selection influences the QF of the current source, (a=1, b=0), then the time complexity is quadratic. On the other hand, for the completely dependent case, the time complexity is $O(|X_i|^N)$, which is exponential as expected from the exhaustive search case. Note that the dual problem, that of minimizing the rate for a given maximum distortion D_{max},

$$\min_{x_0,\ldots,x_{N-1}} R(x_0, \ldots, x_{N-1}) \quad \text{subject to:} \quad D(x_0, \ldots, x_{N-1}) \leq D_{max}, \tag{4.8}$$

can also be solved by the proposed approach using the following relabeling of function names,

$$R(x_0, \ldots, x_{N-1}) \;\leftarrow\; D(x_0, \ldots, x_{N-1}), \tag{4.9}$$

$$D(x_0, \ldots, x_{N-1}) \;\leftarrow\; R(x_0, \ldots, x_{N-1}). \tag{4.10}$$

4.1.5 Example

We use the intra frame encoding mode of H.263 (Test model 4, TMN4 [47]) to give an example of the above concepts. We will also use the results of this example in section 4.3 to compare the minimum total distortion approach with the minimum maximum distortion approach.

We encode the first frame of the quarter common intermediate format (QCIF) color sequence "Mother and Daughter" using TMN4. Each of the 99 16 × 16 blocks is considered a source, for which 31 different quantizers are available. The dependency between the blocks stems from the H.263 mechanism for transmitting the quantizer step sizes which is based on a modified delta modulation scheme. In H.263, the quantizer step size of the current macro block must be within ±2 of the quantizer step size employed for the previous macro block. Then the difference between the quantizer step sizes is entropy coded. This differential pulse code modulation (DPCM) scheme results in a first order dependency between two consecutive blocks, since the operational rate distortion curve of the current block depends on the quantizer selected for the previous block. Therefore the block rate $r_i(x_{i-1}, x_i)$ depends on the current and the previous quantizer, whereas the block distortion $d_i(x_i)$, which we select to be the mean squared error (MSE) between the original and the encoded frame, only depends on the current quantizer selection. The block rate can further be split into a quantizer rate $r_i^Q(x_{i-1}, x_i)$ which reflects the number of bits needed to encode the change in the quantizer step size, and the displaced frame difference (DFD) rate $r_i^{DFD}(x_i)$, which is the number of bits needed to encode the DFD using quantizer x_i. Using these definitions the frame rate can now be expressed as follows,

$$R(x_0, \ldots, x_{N-1}) = \sum_{i=0}^{N-1} r_i(x_{i-1}, x_i) \tag{4.11}$$

$$= \sum_{i=0}^{N-1} r_i^Q(x_{i-1}, x_i) + r_i^{DFD}(x_i). \tag{4.12}$$

Since the MSE is selected as the distortion measure, the frame distortion is simply the sum of the block distortions, that is,

$$D(x_0, \ldots, x_{N-1}) = \sum_{i=0}^{N-1} d_i(x_i). \tag{4.13}$$

The Lagrangian source cost function is therefore of the following form,

$$j_{i,\lambda}(x_{i-1}, x_i) = d_i(x_i) + \lambda r_i(x_{i-1}, x_i), \tag{4.14}$$

and the minimization of the total Lagrangian cost function by dynamic programming can be considered a shortest path problem through the trellis displayed in Fig. 4.1.

To be able to compare the proposed approach with a fixed quantizer encoding of the first frame, we first fix the quantizer step size for all macro blocks to 10. The resulting rate is 18297 bits and the average distortion is 27.8 (MSE). Then we find the optimal quantizers using the above proposed scheme, where the maximum bit rate is set to the number of bits used in the case where the quantizer step size is fixed (R_{max}=18297). The resulting rate is 18431 bits and the average distortion is 27.1. Clearly the proposed scheme was able to find a convex hull solution which is very closed to the desired R_{max} (within one percent) and the average MSE is smaller than the one for the fixed quantizer case. In this example the reduction in distortion is quite small, which shows that a fixed quantizer size for this particular frame and the particular rate is almost optimal. In section 4.3 we will further discuss the results of this approach and compare it to the minimum maximum distortion criterion.

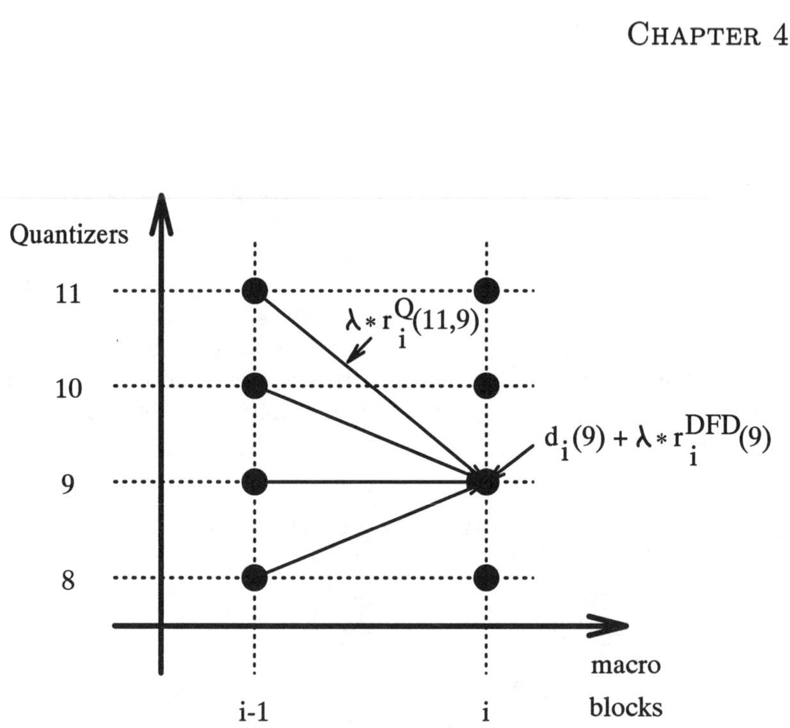

Figure 4.1 Trellis for the image compression example. Each node represents a block quantized with one of the admissible quantizers. The resulting distortion and rate, weighted by the Lagrangian multiplier, are assigned to each node. The transition cost represents the number of bits needed to encode the change in the quantizer step size. The shortest path through the entire trellis results in the best quantizer selection for a given λ.

4.2 VERY FAST CONVEX SEARCH BASED ON A BEZIER CURVE

In this section we introduce a very fast search algorithm for the optimal λ^* needed in the Lagrangian multiplier method [178]. In section 3.3 we have shown that the rate is a non-increasing function of λ and hence the bisection method is a natural choice for finding the optimal λ^*. The bisection method is an iterative scheme and hence for every iteration, the solution to the unconstrained problem needs to be found. Therefore the time complexity of the entire algorithm is directly proportional to the number of iterations. It is therefore desirable to have as few iterations as possible in finding a good λ.

Consider Fig. 4.2 which shows the situation at hand. The proposed iterative search makes use of the fact, that for a continuous rate distortion curve, the derivative is equal to $\frac{-1}{\lambda}$. Assume that the bracketing points $P_0 = (D_0, R_0)$,

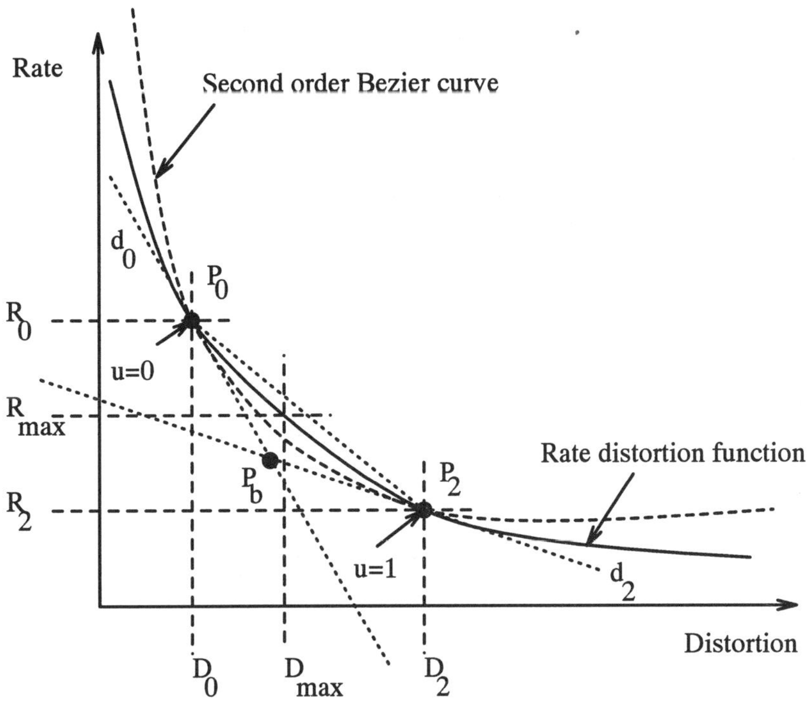

Figure 4.2 Continuous rate distortion curve and the second order Bezier approximation between two bracketing points.

$P_2 = (D_2, R_2)$ and the derivatives at these points, d_0 and d_2, respectively, are given. The goal is to approximate the convex hull between P_0 and P_2 as accurately as possible, such that, this approximation can be used to estimate the derivative d_{max} at R_{max}. This estimate of the derivative $\hat{d}_{max}$ is then used to generate the new $\lambda = \frac{-1}{\hat{d}_{max}}$ for the next iteration. In [164] a fast convex search is used, which is based on the λ theorem presented in section 3.3. The graphical interpretation of the λ theorem is, that given two points on the convex hull and their respective λs, the slope of the line connecting the two points lies between the slope implied by the λs of the points. The search in [164] can be interpreted as approximating the convex hull with a straight line between P_0 and P_2, which is not the best possible approximation since the knowledge of d_0 and d_2 (the negative inverses of the λs) is not used.

A good approximation to the convex hull is a function which goes through the given bracketing points P_0 and P_2 and matches the derivatives d_0 and d_2 at these points. Also the approximation should be convex, differentiable, and in closed form. A parabola would be an ideal choice but it has only three degrees of freedom which is in general not enough to match four independent requirements (P_0, P_2, d_0, d_2). A second order Bezier curve [72] is a parabola which can be tilted in any direction in the plane, hence it has an additional degree of freedom and can therefore match the four independent requirements. A second order Bezier curve is of the following parametric form,

$$P(u) = (1-u)^2 * P_a + 2 * (1-u) * u * P_b + u^2 * P_c. \tag{4.15}$$

By setting $P_a = P_0$ and $P_c = P_2$, two of the four conditions are immediately satisfied, i.e., $P(0) = P_0$ and $P(1) = P_2$. The point P_b can be chosen so that the two derivative conditions are satisfied. The derivative of the second order Bezier curve is,

$$d(u) = \frac{(u-1) * R_0 + (1-2*u) * R_b + u * R_2}{(u-1) * D_0 + (1-2*u) * D_b + u * D_2}. \tag{4.16}$$

Therefore

$$d_0 = d(0) = \frac{R_b - R_0}{D_b - D_0} \tag{4.17}$$

and

$$d_2 = d(1) = \frac{R_2 - R_b}{D_2 - D_b}, \tag{4.18}$$

which leads to $P_b = (D_b, R_b)$ where,

$$D_b = \frac{(R_0 - d_0 * D_0) - (R_2 - d_2 * D_2)}{d_2 - d_0}, \tag{4.19}$$

and

$$R_b = \frac{d_2 * (R_0 - d_0 * D_0) - d_0 * (R_2 - d_2 * D_2)}{d_2 - d_0}. \tag{4.20}$$

As can be seen in Fig. 4.2, P_b is the point where the two tangents of P_0 and P_2 intersect.

Since the goal is to find $\hat{d}_{max}$, the derivative of $P(u)$ has to be evaluated at R_{max}. Hence u_{max} must first be found which leads to $P(u_{max})=(D_{max}, R_{max})$. This can be achieved since $P(u_{max})$ is a second order function of u_{max} for which the quadratic equation can be solved explicitly. Therefore, from Eq. (4.15) we obtain,

$$u_{max}^{\pm} = -\frac{R_b - R_0}{(R_2 - R_b) - (R_b - R_0)}$$
$$\pm\sqrt{\frac{(R_b - R_0)^2}{((R_2 - R_b) - (R_b - R_0))^2} + \frac{R_{max} - R_0}{(R_2 - R_b) - (R_b - R_0)}}, \quad (4.21)$$

where u_{max} is picked such that $1 \geq u_{max} \geq 0$. In the case where $(R_2 - R_b) - (R_b - R_0)$ equals zero, u_{max} equals $(R_{max} - R_0)/(2 * (R_b - R_0))$. Hence the estimate of the derivative at R_{max}, is given by $\hat{d}_{max} = d(u_{max})$.

The search for the optimal λ^* follows the outline of the bisection method, but instead of setting the new λ equal to the average of the old λs, the above scheme is used to find the new λ. In our experiments, the iterative search which is based on the second order Bezier approximation exhibits a very fast convergence and hence we call it the very fast convex search.

4.3 OPTIMAL BIT ALLOCATION FOR DEPENDENT QUANTIZERS USING THE MINIMUM MAXIMUM DISTORTION CRITERION

In this section we introduce an optimal bit allocation scheme for dependent quantizers for the minimum maximum distortion criterion. First we show how minimizing the bit rate for a given maximum distortion can be achieved in a dependent coding framework using dynamic programming. Then we employ an iterative algorithm to minimize the maximum total distortion for a given bit rate. This iterative algorithm invokes the dynamic programming scheme and we prove that it converges to the optimal solution.

As we have pointed out before, there exists an inherent tradeoff between the rate and the distortion of a lossy compression scheme. One common approach to mathematically formulate this relationship is to minimize the total distortion for a given maximum bit rate, or vice versa, to minimize the bit rate for a given maximum distortion. The philosophy behind this approach is that if the average (total) distortion is minimized then, in the long run, the best quality is obtained. We have shown in section 3.3 that the Lagrangian multiplier method is well suited for these kind of constrained optimization problems. It is interesting to notice that with this popular approach, a large variability among the different source distortions is possible. When the sources are consecutive in time or space, such as different frames of a sequence or different blocks in a frame, this variability in quality can be very disturbing and the perceived quality is fairly low even though the average distortion is minimized.

A different approach to formalize the relationship between the rate and the distortion is the minimum maximum distortion approach, where the goal is to minimize the maximum distortion for a given bit rate, or vice versa, to minimize the bit rate for a given maximum distortion. The philosophy behind this approach is that by minimizing the maximum distortion, no single source distortion will be extremely bad and hence the overall quality will be quite constant. In fact, the minimum maximum distortion approach is a natural criterion to be selected when the goal is to achieve an almost constant distortion which is as small as possible for the available bit rate.

As we have pointed out in section 3.2.1 the minimum maximum distortion problem for independent quantizers has been studied in [105] and a simple algorithm has been proposed. In this section we introduce a scheme which finds the optimal solution for the minimum maximum distortion criterion for depended quantizers, where the dependency can be arbitrarily large.

4.3.1 Notation and assumptions

We assume that the rate r_i and the distortion d_i of a source i depend on the quantizer selections of neighboring sources in a neighborhood defined by two non-negative integers a and b, that is,

$$R(x_0, \ldots, x_{N-1}) = \sum_{i=0}^{N-1} r_i(x_{i-a}, \ldots, x_{i+b}), \tag{4.22}$$

$$D(x_0, \ldots, x_{N-1}) = \max_{i \in [0, \ldots, N-1]} \{d_i(x_{i-a}, \ldots, x_{i+b})\}, \tag{4.23}$$

where x_i is the quantizer of source i selected from the set of admissible quantizers X_i for source i and N is the total number of sources. As in section 3.4 the quantizers $x_{-a}, \ldots, x_{-1}$ and $x_N, \ldots, x_{N-1+b}$ are the boundary parameters and can be set to any desired value.

4.3.2 The minimum rate case

First we consider the case where the goal is to find the minimum rate for a given maximum distortion D_{max}. This can be formulated as follows,

$$\min_{x_0,\ldots,x_{N-1}} R(x_0,\ldots,x_{N-1}) \quad \text{subject to:} \quad D(x_0,\ldots,x_{N-1}) \leq D_{max}. \quad (4.24)$$

The key observation for the proposed optimal bit allocation procedure is that the maximum distortion D_{max} is a constraint which applies to each source distortion $d_i(x_{i-a},\ldots,x_{i+b})$, as is clear from Eq. (4.23). Therefore any quantizer selection which results in a single source distortion being larger than D_{max} cannot belong to the optimal quantizer sequence. Of the remaining quantizer sequences, the one which results in the smallest rate is the optimal solution to the above problem. These two concepts can be combined using the following definitions,

$$g_i(x_{i-a},\ldots,x_{i+b}) = \begin{cases} \infty & : \ d_i(x_{i-a},\ldots,x_{i+b}) > D_{max} \\ r_i(x_{i-a},\ldots,x_{i+b}) & : \ d_i(x_{i-a},\ldots,x_{i+b}) \leq D_{max} \end{cases}, \quad (4.25)$$

and

$$G(x_0,\ldots,x_{N-1}) = \sum_{i=0}^{N-1} g_i(x_{i-a},\ldots,x_{i+b}). \quad (4.26)$$

Finding the minimum of the objective function in Eq. (4.26) is equivalent to the original problem

$$R^*(D_{max}) = \min_{x_0,\ldots,x_{N-1}} R(x_0,\ldots,x_{N-1}) \ \text{subject to:} \ D(x_0,\ldots,x_{N-1}) \leq D_{max} \quad (4.27)$$

where $R^*(D_{max})$ is the smallest rate for a given maximum distortion D_{max}. We have shown in section 3.4 that dynamic programming can be used to solve the following problem optimally and efficiently,

$$\min_{x_0,\ldots,x_{N-1}} G(x_0,\ldots,x_{N-1}), \quad (4.28)$$

when $G(x_0,\ldots,x_{N-1})$ has the structure given in Eq. (4.26). Hence DP can find the optimal solution to problem (4.27). If the optimal solution $[x_0^*,\ldots,x_{N-1}^*]$ result in an infinite rate $R^*(D_{max})$, then there exits no solution which can satisfy the maximum distortion requirement for each source. When the rate for the optimal solution is finite, then the solution optimally selects the admissible quantizers such that the rate $R^*(D_{max})$ is minimized for a given maximum distortion D_{max}.

4.3.3 The minimum distortion case

In the minimum distortion case, the maximum number of bits (R_{max}) is given and the goal is to select the quantizers in such a fashion that the resulting maximum distortion is as small as possible. Mathematically this can be expressed as follows,

$$\min_{x_0,\ldots,x_{N-1}} D(x_0,\ldots,x_{N-1}) \quad \text{subject to:} \quad R(x_0,\ldots,x_{N-1}) \le R_{max}. \quad (4.29)$$

The main difference to the minimum rate problem is the fact that the maximum rate R_{max} is not an upper limit for each source, but for the sum of the sources. Also this is a minimum maximum problem, since the distortion is defined as the maximum over all source distortions.

The proposed optimal bit allocation algorithm for the minimum distortion case is based on the fact that we can optimally solve the minimum rate case. In other words, for every given D_{max} we can find the quantizer sequence which results in $R^*(D_{max})$, the minimum rate for encoding the combined sources, where each source distortion has to be below the maximum distortion D_{max}. We use the following lemma to formulate an iterative procedure to find the optimal solution for the minimum distortion problem.

Lemma 1 *$R^*(D_{max})$ is a non-increasing function of D_{max}.*

Proof: Let $D^2_{max} \ge D^1_{max}$, let $[{}^1x^*_0,\ldots,{}^1x^*_{N-1}]$ be the optimal solution of Eq. (4.27) for $D_{max}=D^1_{max}$, and let $[{}^2x^*_0,\ldots,{}^2x^*_{N-1}]$ be the optimal solution of Eq. (4.27) for $D_{max}=D^2_{max}$. Since $D^1_{max} \le D^2_{max}$, $[{}^1x^*_0,\ldots,{}^1x^*_{N-1}]$ is a possible solution of Eq. (4.27) for $D_{max}=D^2_{max}$, using $R^*(D^1_{max})$ bits. Since $[{}^2x^*_0,\ldots,{}^2x^*_{N-1}]$ is the optimal solution of Eq. (4.27) for $D_{max}=D^2_{max}$, it follows that $R^*(D^2_{max}) \le R^*(D^1_{max})$. ■

The above lemma is intuitively clear since it simply states that if a greater maximum error is permissible, then we should be able to encode the sources with a smaller number of bits. Note that even though this seems obvious, this only holds because we can solve the minimum rate case optimally.

Having shown that $R^*(D_{max})$ is a non-increasing function, we can use the bisection method [59] to find the optimal D^*_{max} such that $R^*(D^*_{max}) = R_{max}$, which solves the minimum distortion problem of Eq. (4.29). The bisection method starts with two points $(D^l_{max}, R^*(D^l_{max}))$ and $(D^u_{max}, R^*(D^u_{max}))$ which bracket

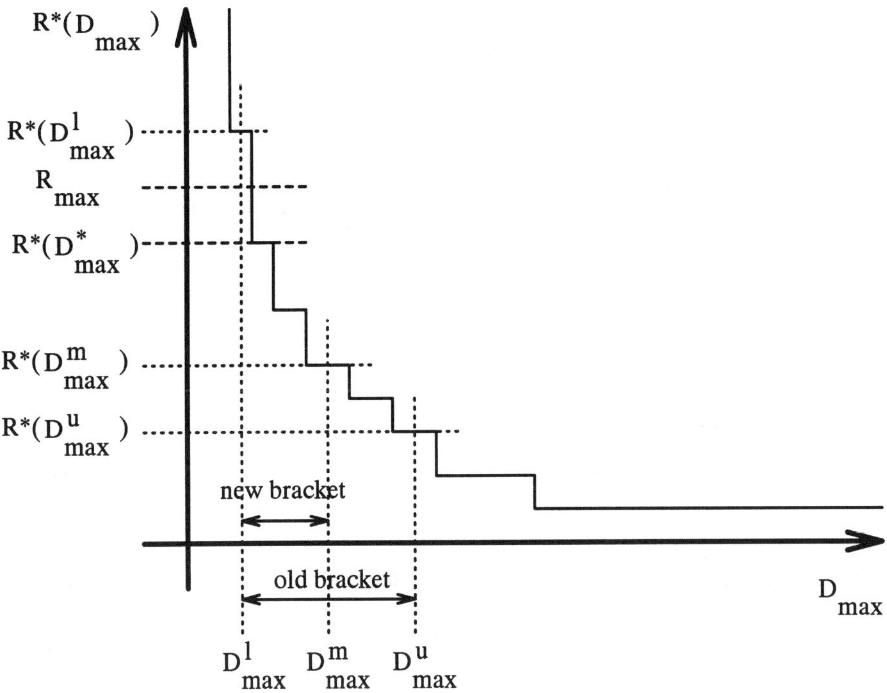

Figure 4.3 The $R^*(D_{max})$ function, which is a non-increasing function exhibiting a staircase characteristic. The selected R_{max} falls onto a discontinuity and therefore the optimal solution is of the form $R^*(D^*_{max}) < R_{max}$, instead of $R^*(D^*_{max}) = R_{max}$.

the optimal solution (see Fig. 4.3). Then a middle point $(D^m_{max}, R^*(D^m_{max}))$ is found by invoking the minimum rate algorithm for,

$$D_{max} = D^m_{max} = (D^l_{max} + D^u_{max})/2. \tag{4.30}$$

The new bracketing points of the optimal solution are then the middle point and one of the original points which results in a bracket which includes the optimal solution. This procedure is then iterated until the optimal solution is found or the bracket is small enough for the purpose at hand.

Since this is a discrete optimization problem, the function $R^*(D_{max})$ is not continuous and exhibits a staircase characteristic (see Fig. 4.3). This implies that there might not exist a D^*_{max} such that $R^*(D^*_{max}) = R_{max}$. In that case, the proposed algorithm will still find the optimal solution, which is of the form $R^*(D^*_{max}) < R_{max}$, but only after an infinite number of iterations. This is true, since after every iteration, the length of the interval which contains the optimal solution is cut in half. Hence after a any finite number of iterations, the interval still contains infinitely many possible solutions, but after an infinite number of iterations, the length of the interval is zero and the optimal solution is found. In practice, if we have not found a D_{max} such that $R^*(D_{max}) = R_{max}$ after a given maximum number of iterations, we terminate the algorithm.

4.3.4 Example

In this section we present a comparison between the minimum total (or average) distortion approach and the minimum maximum distortion approach. The dependent image coding scheme we use for this example is the intra frame scheme employed in TMN4 [47], which is the test model four of the H.263 standard.

We encode the first frame of the QCIF color sequence "Mother and Daughter" using this scheme. We use the H.263 mechanism for transmitting the quantizer step sizes which is based on a modified delta modulation scheme. In H.263, the quantizer step size of the current macro block must be within ± 2 of the quantizer step size employed for the previous macro block. Then the difference between the quantizer step sizes is entropy coded. This DPCM scheme results in a first order dependency between two consecutive blocks, since the operational rate distortion curve of the current block depends on the quantizer selected for the previous block.

First we fix the quantizer step size for all macro blocks to 10. The resulting rate and distortion are listed in Table 4.1. Note that the MSE of the luminance (Y) channel is used as the distortion measure. In the case where we minimize the total distortion subject to a maximum bit rate, the problem can be formulated using the Lagrangian multiplier method. The solution of the relaxed (unconstrained) problem can then be found using dynamic programming. We discussed this approach in section 4.1. Again, the resulting rate and distortion are in Table 4.1 and so are the results for the proposed minimum maximum distortion scheme. Note the similarity between the minimum total distortion scheme and the minimum maximum distortion approach. For the former, an iteration (for example, the very fast convex search presented in section 4.2) is need to find the optimal tradeoff parameter λ, where for each iteration the relaxed problem is solved by dynamic programming. For the latter, bisection,

	Rate	mean MSE	min MSE	max MSE	std MSE
fixed Q=10	18297	27.8	0.6	66.8	17.5
min total Dist.	18431	27.1	0.6	65.0	17.3
min max Dist.	18293	29.9	0.6	46.2	15.9

Table 4.1 Comparison between the minimum total distortion and the minimum maximum distortion paradigm, using the first frame of the QCIF (176 × 144) "Mother and Daughter" sequence. Each of the 99 macro blocks results in a particular MSE, and the mean MSE is the mean of these 99 MSEs. The same holds for the minimum, maximum and standard deviation column.

also an iterative scheme, is employed, where for each iteration, the minimum rate problem is solved optimally using dynamic programming. One of the main differences is that the Lagrangian approach can only find solutions which belong to the convex hull, whereas the proposed minimum maximum distortion scheme will always find the optimal solution.

In Table 4.1 the statistics for these three different approaches are given. The table reflects the different paradigms behind the minimum total distortion approach and the minimum maximum distortion approach. Clearly the minimum maximum distortion approach results in a higher average distortion, on the other hand, its maximum distortion is much lower than the maximum distortion of the other two schemes. This is also reflected in the smaller standard deviation. In other words, the minimum maximum distortion approach results in an encoded image, with more uniform quality than the one resulting from the minimum total distortion approach.

In Fig. 4.4 the MSE per macro block for the three implementations is shown and in Fig. 4.5 the corresponding quantizer selections are displayed. It is interesting to notice in Fig. 4.5 that there are quite a few blocks where the quantizers are the same for both optimal schemes. These blocks tend to coincide with the blocks where the MSE (see Fig. 4.4) is very small, i.e., blocks with no high frequency components. It is clear from Fig. 4.4 that the minimum maximum distortion scheme results in a more even quality for the entire frame than the minimum total distortion approach. In Fig. 4.6 the resulting Y channels of the three approaches are compared to the original image. The compression ratio for the three compressed images is approximately 11 and it is clear that even though the block MSEs are quite different, the visual quality of the three images in this printed reproduction is about the same.

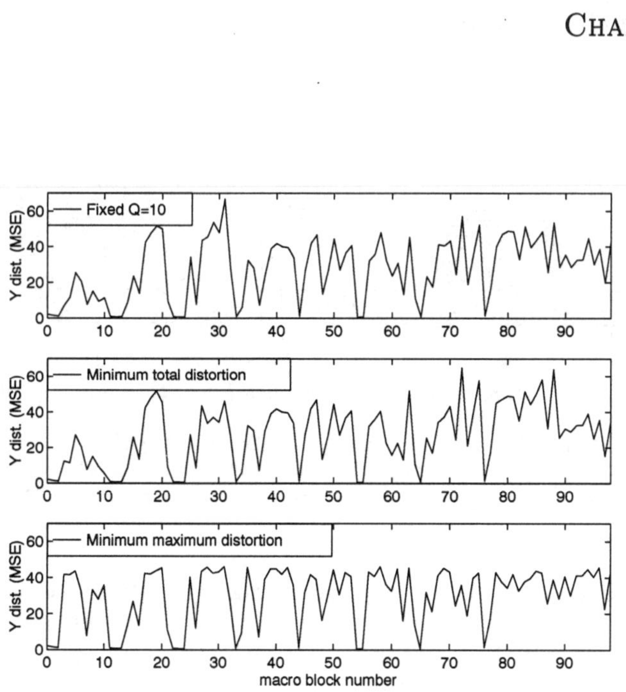

Figure 4.4 The macro block MSE of the luminance channel is displayed. In the first row, the MSE for a fixed quantizer step size of Q=10 is shown. In the second row, the MSE for the minimum total distortion approach is displayed and in the third row, the MSE for the minimum maximum distortion approach is shown.

4.4 OPTIMAL SCANNING PATH FOR A QUAD-TREE DECOMPOSITION

We introduced in section 2.3 the quad-tree (QT) data structure as an efficient representation of a variable block size frame segmentation. In this section we consider the problem of finding an optimal scanning path for an arbitrary QT decomposition of the frame. A scanning path is a rule which defines in what order the different blocks of a frame are visited. Clearly if the QT decomposition changes from frame to frame, each frame requires a different scanning path. We propose a procedure for inferring a scanning path from a given QT decomposition. For this procedure, we derive a necessary and sufficient condition which guarantees that consecutive blocks along the scanning path are always neighboring blocks.

When a DPCM scheme along the scanning path is employed, the fact that consecutive blocks along the scanning path are neighboring blocks, is very important. This is because the spatial correlation among neighboring blocks is

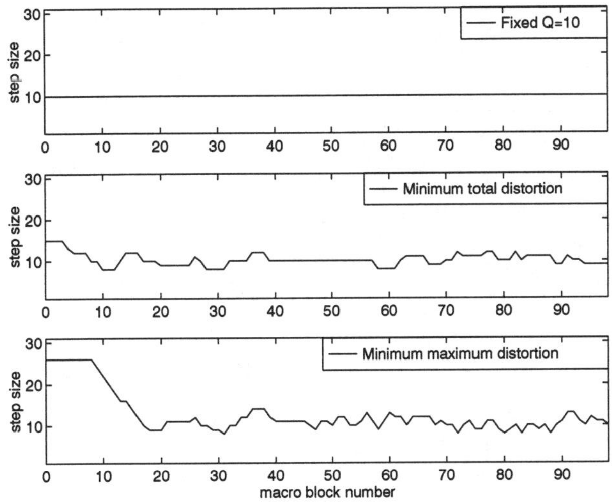

Figure 4.5 The macro block quantizer step sizes are displayed. In the first row, the fixed quantizer step size of Q=10 is shown. In the second row, the step sizes for the minimum total distortion approach are displayed and in the third row, the step sizes for the minimum maximum distortion approach are shown.

high, and the higher the correlation, the more efficient the DPCM scheme is. In later chapters of the monograph, we will employ a QT-based scheme to encode the displacement vector field (DVF), where the DVF is first order DPCM encoded along the scanning path.

4.4.1 Quad-tree notation

We assume that the frame is segmented by a QT, as in Fig. 4.7, where the black curve indicates the feature of interest. As can be seen in this figure, the QT data structure decomposes a $2^N \times 2^N$ image (or block of an image) down to blocks of size $2^{n_0} \times 2^{n_0}$. This decomposition results in an $(N - n_0 + 1)$-level hierarchy ($0 \leq n_0 \leq N$), where all blocks at level n ($n_0 \leq n \leq N$) have size $2^n \times 2^n$. This structure corresponds to an inverted tree, where each $2^n \times 2^n$ block (called a *tree node*) can either be a *leaf*, i.e., it is not further subdivided, or can branch into four $2^{n-1} \times 2^{n-1}$ blocks, each a *child* node. The tree can be

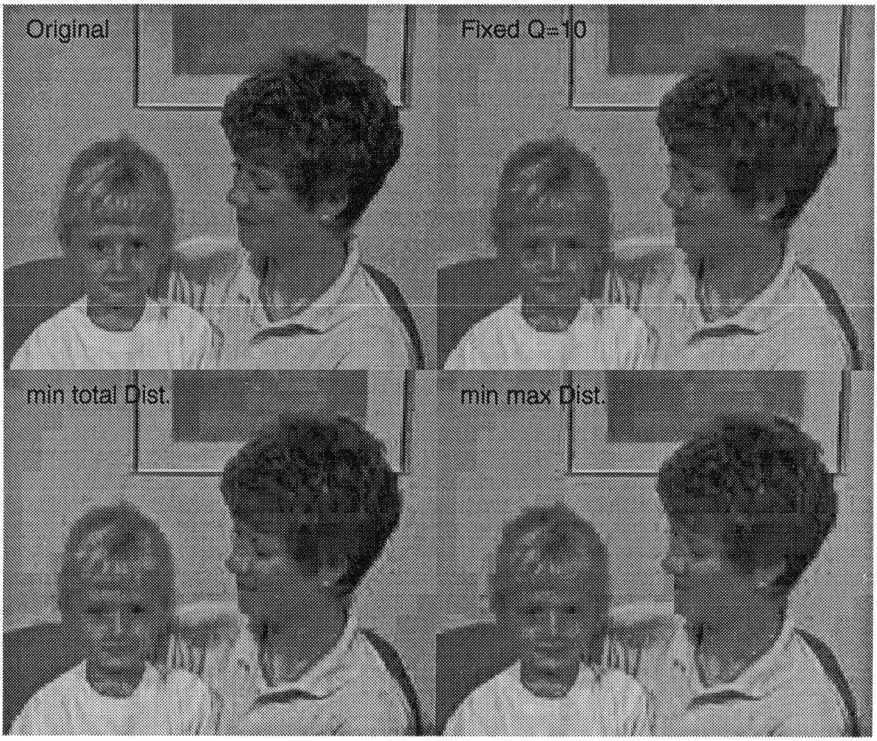

Figure 4.6 Comparison between Y channels of the three approaches.

represented by a series of bits that indicate termination by a leaf with a "0" and branching into child nodes with a "1" (see Fig. 4.8).

Let $b_{l,i}$ be the block i at level l, and the children of this block are therefore $b_{l-1,4*i+j}$, where $j \in [0,1,2,3]$. The complete tree is denoted by $\mathcal{T}$ and a tree node is identified by the ordered pair (l,i), where again l is the level and i the number within that level (see Fig. 4.9). This ordered pair is called the index of the tree node. Each leaf of $\mathcal{T}$ represents a particular block in the segmented frame. For future convenience, let the leafs be numbered from zero to the total number of leafs in the QT ($N_{\mathcal{T}}$) minus one, from left-to-right and hence in increasing order of the measure $4^{l-n_0} * i$ (this ordering of the leafs is indicated by italic numbers in Fig. 4.9).

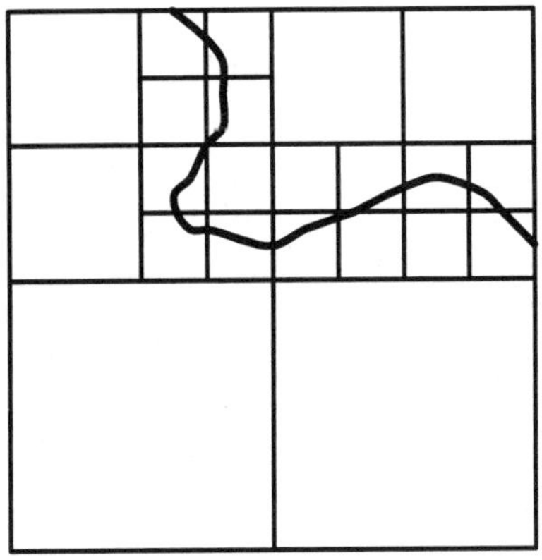

Figure 4.7 Frame segmented by a quad-tree.

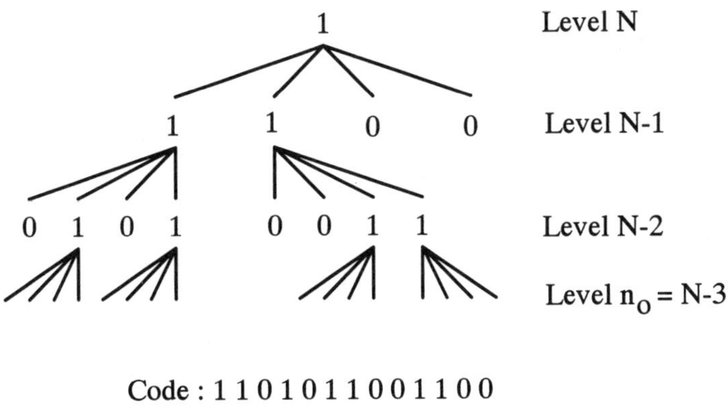

Figure 4.8 Quad-tree representation of the frame.

4.4.2 Procedure and condition

By decomposing a $2^N \times 2^N$ image block using a QT into sub-blocks of different sizes ($2^n \times 2^n, n_0 \leq n \leq N$), the problem of finding a good scanning path arises, since we assume that a first order DPCM scheme along the scanning path is employed. Note that a different scanning path is necessary for each different QT decomposition of the frame. Ideally, the scanning path should

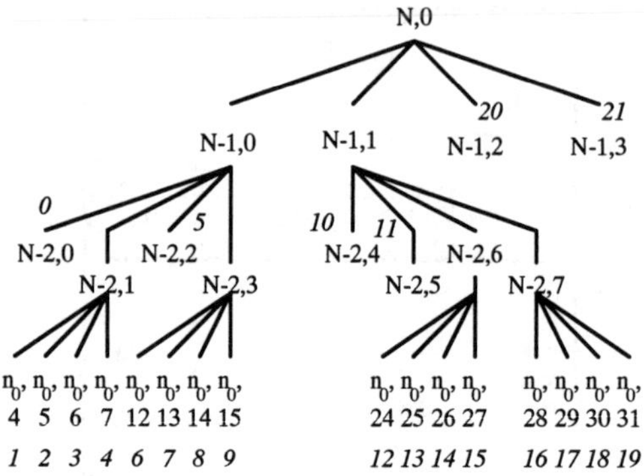

Figure 4.9 Quad-tree notation. Each node is identified by the ordered pair (l, i), where l is the level and i the number within that level. Also the leafs are numbered from zero to the total number of leafs in the QT minus one.

be inferred from the QT decomposition, so that no additional information is required, which in a coding scheme, would have to be sent as overhead.

Clearly, the traditional raster scan (left to right, top to bottom), which is the same for every frame, cannot be applied to variable block sizes since the QT segmentation of the scene changes from frame to frame. To simplify the search for a good scanning path, we require that child blocks, which belong to the same parent block, are scanned in sequence.

Since the QT decomposition is recursive in nature, a recursive version of the raster scan is commonly used to scan the QT in the predefined raster scan pattern which is the same for each block. In Fig. 4.10 an example of this recursive raster scan is shown. Fig. 4.10a through Fig. 4.10c show the scanning path at different levels and in Fig. 4.10d the resulting overall scanning path is displayed. Since the goal is to use a first order DPCM scheme along the scanning path, the correlation between successive blocks, and therefore the efficiency of the DPCM, strongly depends on the selected scanning path. The following definition explains how we interpret the concept of a "good" scanning path.

Definition 1 *A scanning path is optimal if it connects only neighboring blocks, i.e., blocks which share an edge, and if it visits each block once and only once.*

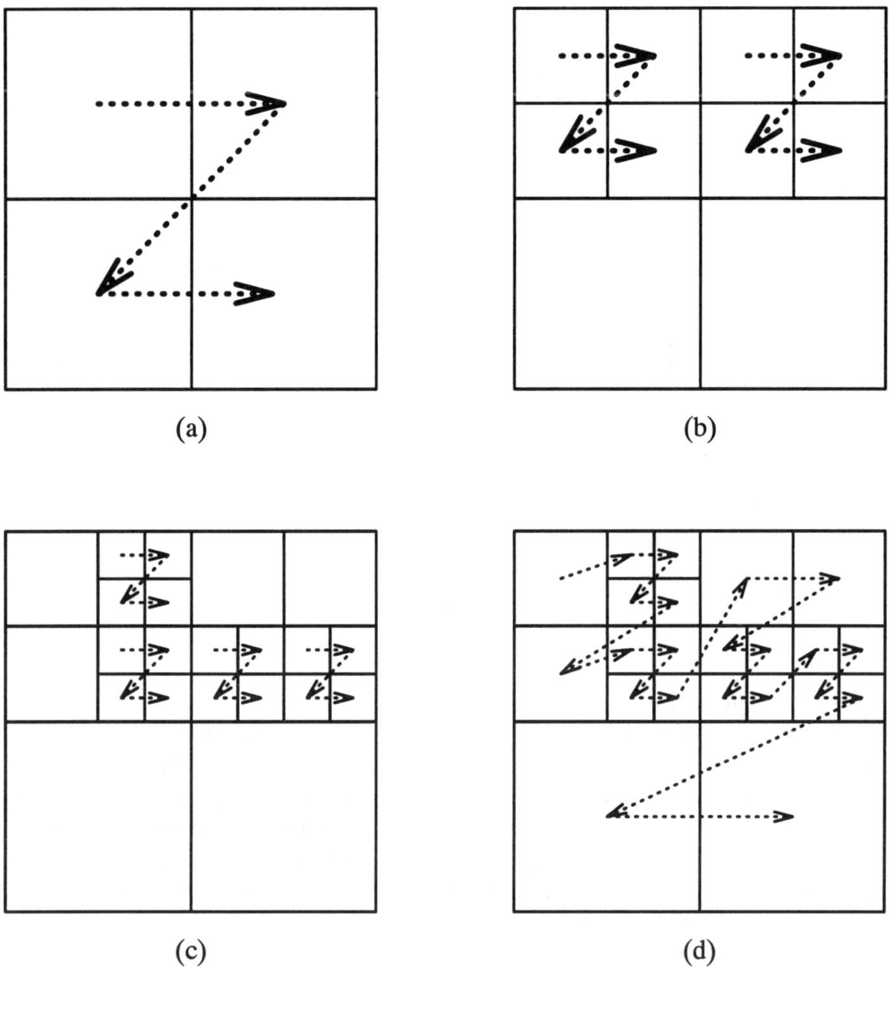

Figure 4.10 Recursive raster scan for a QT decomposition. Figures (a) through (c) show the raster scan at different levels of the QT and figure (d) shows the resulting overall scanning path.

The motivation behind this definition comes from the observation that an optimal scanning path will lead to highly correlated successive blocks, since such blocks will always be neighboring blocks. To illustrate this concept, Fig. 4.11 shows an optimal and a suboptimal scanning path. Note that the recursive

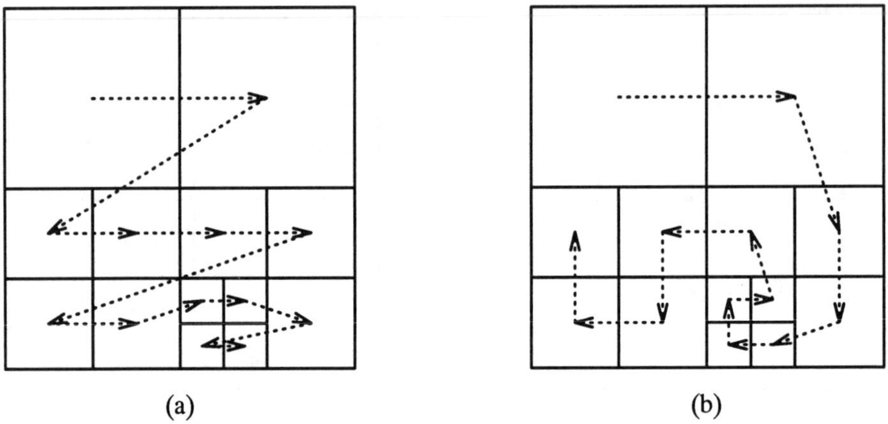

Figure 4.11 Scanning paths: (a) suboptimal raster scan-type path; (b) optimal scanning path

raster scan, which has been traditionally used in the QT decomposition, is suboptimal, as can be seen from Fig. 4.10a.

Since the segmentation changes from frame to frame, no fixed scanning path can be employed. On the other hand, no additional information should have to be transmitted to indicate which scanning path was used. Therefore, a good scanning path must be inferable from the transmitted QT and it has to be optimal in the above defined sense. In the following section we introduce a procedure to infer a scanning path from the transmitted QT decomposition and derive a necessary and sufficient condition for that procedure, so that the resulting scanning path is always optimal.

Procedure

Assume that the frame is completely segmented, as in Fig. 4.12, into blocks of the smallest size, i.e., $2^{n_0} \times 2^{n_0}$. The QT representation of this segmentation is a complete tree, where all the leafs are of level n_0. Further assume that the scanning path for this decomposition is known to the encoder and decoder. Let the scanning path for any other QT decomposition be defined recursively, by merging four consecutive blocks along the scanning path at the lower level to form the blocks at the higher level. Note that with this definition, the blocks do not need to be square. An example of this recursive definition is given in

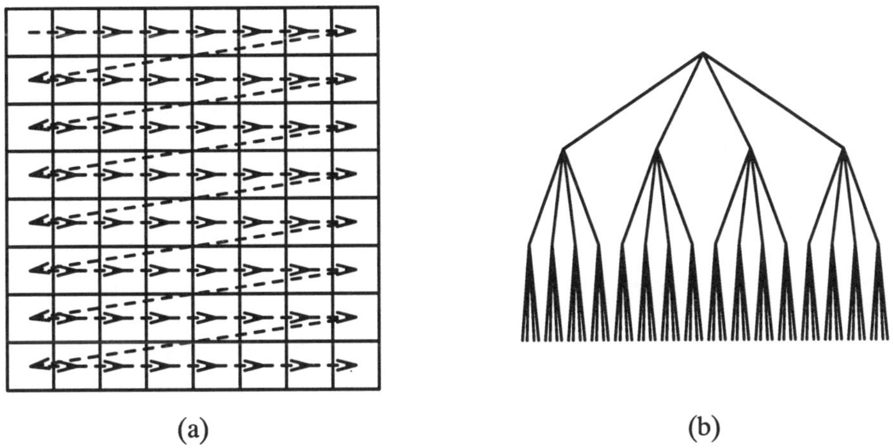

Figure 4.12 Completely decomposed quad-tree. Level n_0 is scanned with a raster scan.

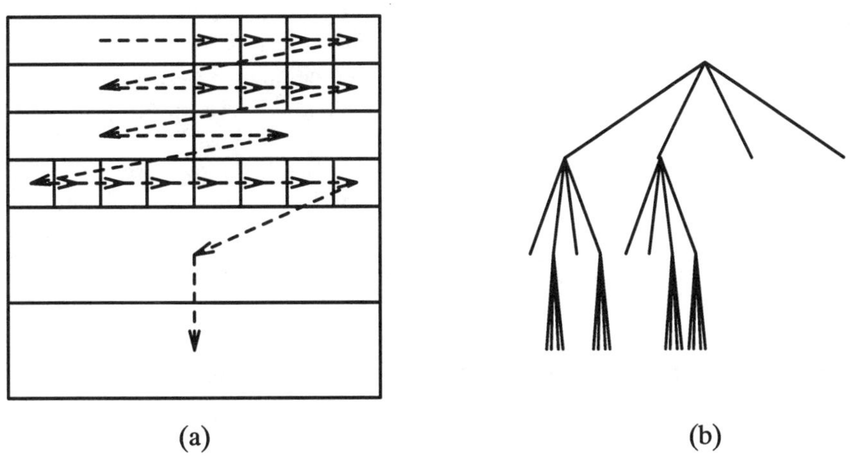

Figure 4.13 Recursive definition of the scanning path.

Fig. 4.13, where the level n_0 scanning path is the raster scan as displayed in Fig. 4.12.

Clearly this recursive definition of the scanning path results in a scanning path that is inferable from the QT decomposition and the scanning path of the fully decomposed QT. The following lemma establishes a necessary and sufficient condition on the level n_0 scanning path, so that, using the above recursive definition of the scanning path for any QT decomposition, the resulting scanning path is always optimal.

Necessary and sufficient condition

Lemma 2 *If the level n_0 scanning path is optimal, and the above recursive procedure is used for the generation of a scanning path for any QT decomposition, then the resulting scanning path is optimal.*

Proof: necessary condition: if it is violated, then a completely decomposed tree (down to level n_0) would not result in an optimal scanning path since the level n_0 scanning path is not optimal.

Proof: sufficient condition (by induction): The level n_0 scanning path is optimal. Consider an optimal scanning path. Pick any connected sub path visiting four blocks. Since the overall scanning path is optimal, the sub path is optimal too. By merging the four blocks of the sub path into a single block, the neighboring blocks of the four merged blocks are also neighboring blocks of the new single block. Hence the new scanning path is optimal. ■

From the above lemma it is clear how to correct the scanning path in Fig. 4.12 so that the scanning path in Fig. 4.13 is optimal. One has to make sure that the level n_0 scanning path is optimal. A slightly changed raster scan displayed in Fig. 4.14 results in the optimal scanning path in Fig. 4.15.

4.4.3 Optimal scanning path based on a Hilbert curve

In the previous section we have established a procedure and a necessary and sufficient condition for generating optimal scanning paths. Clearly there are many different optimal scanning paths, since there are many different optimal level n_0 scanning paths. One problem with the previous procedure is that an

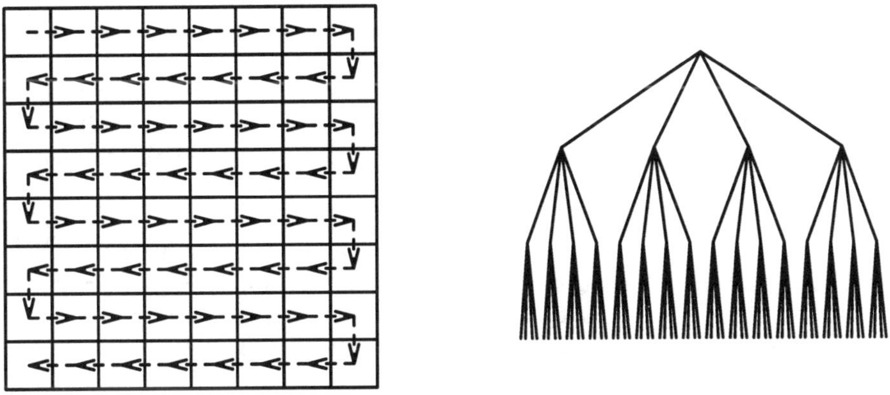

(a) (b)

Figure 4.14 Corrected raster scan.

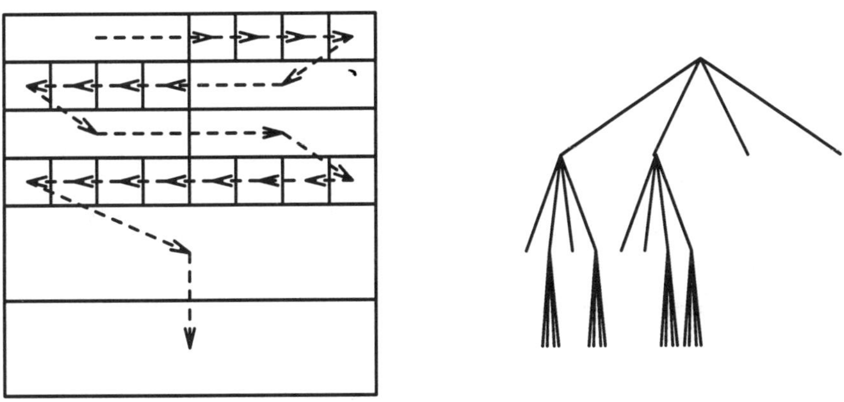

(a) (b)

Figure 4.15 Optimal scanning path.

optimal level n_0 scanning path does not guarantee that the resulting blocks are squares for any QT decomposition. In this section we propose an optimal scanning path which results in square blocks, and hence in the kind of QT decomposition displayed in Fig. 4.7.

The optimal scanning path we developed is based on a Hilbert curve. Note that Hilbert curves are also called Peano curves [114]. A Hilbert curve has a certain order D and the first order Hilbert curve can have one of four orientations, denoted by 0, 1, 2 and 3. Figure 4.16 shows the four first order Hilbert curves with the associated second order Hilbert curves drawn beneath it. Note that each second order Hilbert curve consists of four first order Hilbert curves connected by a dotted line and scaled by a factor of $\frac{1}{2}$.

This relationship between first and second order Hilbert curves can be generalized using the following proposed algorithm. This algorithm generates a Hilbert curve of order N and shows the close relationship between a QT and a Hilbert curve. Let $O_{l,i} \in [0, 1, 2, 3]$ be the orientation (scanning order of the blocks $b_{l-1,4*i+j}, j \in [0, 1, 2, 3]$) of block $b_{l,i}$ and let $(\cdot)_4$ denote the modulo 4 operation. The recursion needs to be initialized by the desired orientation of the first order Hilbert curve, say $O_{N,0} = 0$ (Figure (a) in Fig. 4.16). Then we propose to create a Hilbert curve of order N in a Top-Down fashion by calling the function "orient(l,i)" below with the following parameters: orient(N,0).

orient(l, i)
  $O_{l-1,4*i+0} = (5 - O_{l,i})_4;$
  $O_{l-1,4*i+1} = O_{l,i};$
  $O_{l-1,4*i+2} = O_{l,i};$
  $O_{l-1,4*i+3} = 3 - O_{l,i};$
  if $(l - 1 > 1)$
    orient$(l - 1, 4 * i + 0);$
    orient$(l - 1, 4 * i + 1);$
    orient$(l - 1, 4 * i + 2);$
    orient$(l - 1, 4 * i + 3);$

The resulting orientations at each QT level l, $(N \geq l \geq 1)$ correspond to a Hilbert curve of order $N - l + 1$, hence at level one, the orientations form a Hilbert curve of order N. In Fig. 4.17 the recursive generation of a third order Hilbert curve is shown. This curve was generated by using the following function call: orient(3,0).

Hilbert curves have been used in image and video processing as scanning paths on the pixel level in the luminance domain for lossless coding [159] and lossy coding [119]. They have also been used as a scanning path for the coefficients in

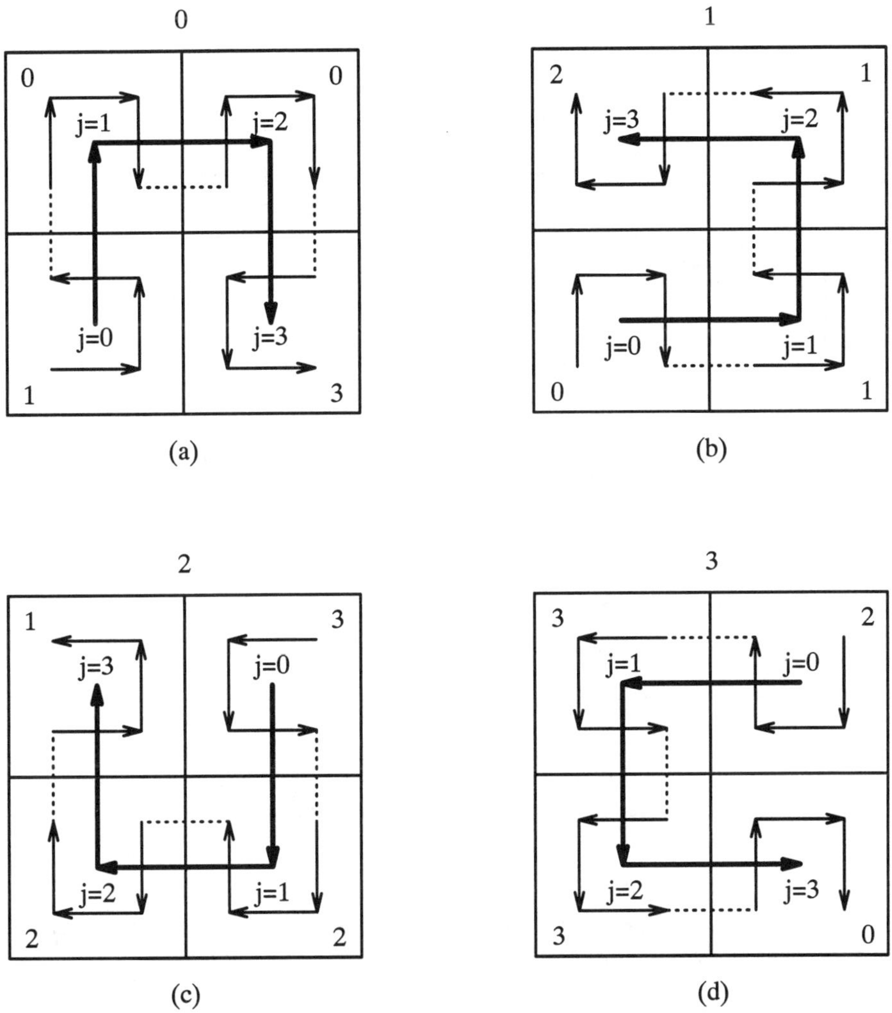

Figure 4.16 The four possible first order Hilbert curves (-) with the respective second order Hilbert curves underneath (-). The orientations $O_{l,i}$ are shown on top of the squares and the orientations $O_{l-1,4*i+j}$ are shown in the corners of the smaller squares.

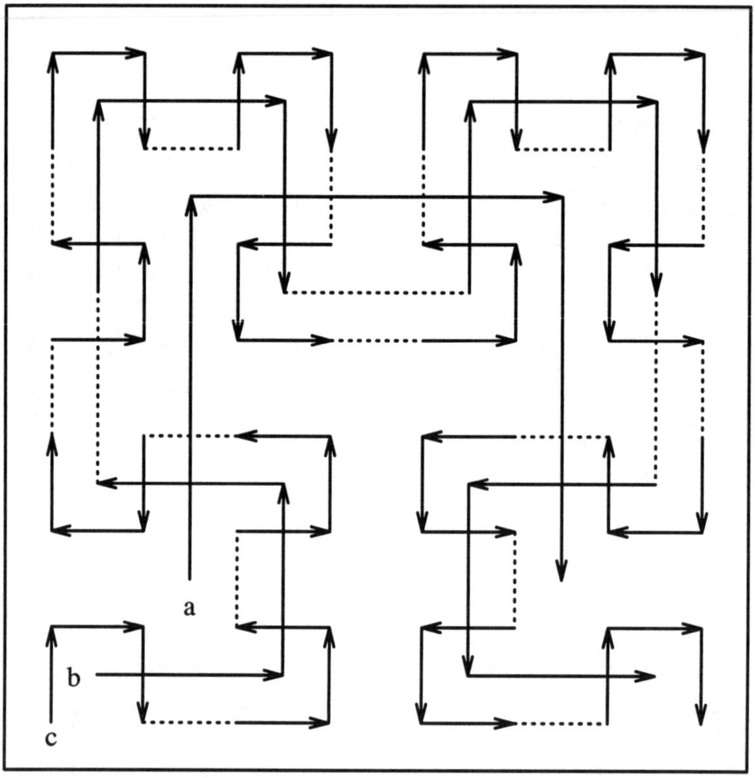

Figure 4.17 Recursive Hilbert curve generation: a) first order Hilbert curve of orientation 0, b) second order Hilbert curve, c) third order Hilbert curve

the transform domain [39]. In all these cases, the fact that an image, which is scanned according to a Hilbert curve, creates an one-dimensional representation of the image which is more correlated than a raster scan, is exploited.

In [119] it is noted that the recursive form of the Hilbert curve is very similar to a QT decomposition of the image down to the pixel level. It is further noted that a Hilbert curve scan differs from a standard ordering in that different orientations are used in successive quadrants. The above procedure formalizes this strong relationship between a QT and a Hilbert curve.

The Hilbert curves are said to be "space filling." Hilbert proved that as the order tends towards infinity, the infinitely thin, continuous line of the Hilbert curve passes trough every point of the original square. Mandelbrot's [114] definition of a fractal states that a curve is a fractal if its Hausdorff dimension exceeds its topological dimension. Based on this definition, as the order tends towards infinity, the Hilbert curve is a fractal since its Hausdorff dimension equals 2 [175] and therefore exceeds its topological dimension which is 1 as for every curve.

Having defined the relationship between a Hilbert curve and a QT and discussed some properties of the Hilbert curve, it is clear that when the level n_0 scanning path is a Hilbert curve of order $N - n_0$, then the resulting scanning path is optimal, regardless of the QT decomposition and the resulting blocks are square. Besides resulting in square blocks, the Hilbert scan has also the advantage that it is spatially non-disruptive and hence tends to preserve local pixel correlations better than a raster scan.

4.5 OPTIMAL QUAD-TREE DECOMPOSITION WITH LEAF DEPENDENCIES

In this section we propose an algorithm which finds the optimal QT decomposition in the rate distortion sense for a QT with leaf dependencies. We will derive the algorithm using the minimum total distortion criterion and then point out in section 4.5.8 how the approach needs to be adjusted for the minimum maximum distortion approach.

4.5.1 Introduction

The underlying problem is the encoding of a given frame by a variable block size scheme, where the block size is encoded using a QT, and each block is encoded by one of the admissible quantizers for that block. Several different schemes which are based on this structure have been recently proposed [195, 196, 210, 234, 192]. All these approaches employ heuristic rules for finding the quantizers and the QT decomposition, whereas the following two approaches address the problem of optimal QT decomposition and quantizer selection.

In [29] a general scheme for tree-structured source coding and modeling is presented and one of the applications mentioned is the optimal QT decomposition in the rate distortion sense. In [198, 200] a scheme is proposed, which is based on the Lagrangian multiplier method, to find the best QT decomposition in the rate distortion sense. We discussed both approaches in section 3.2.1. Both schemes assume that the nodes of the QT are independent and hence cannot be applied when there is a dependency between the nodes.

A dependency between the nodes arises from the use of predictive coding such as DPCM. For example, in chapter 5 we represent the inhomogeneous DVF using a QT, where the motion vectors of the QT leafs are encoded using first order DPCM. In this section, we use the following example of a mean value QT decomposition to illustrate the algorithm. A mean value QT decomposition segments the image into different blocks of different sizes using a QT, where for each block only the mean value is transmitted. Clearly one can quantize and encode each mean value separately, which results in a rate and a distortion pair for each QT node, which is independent of all other QT nodes. Then the problem is to find the best QT decomposition for a given bit rate, so that the distortion is minimized. As mentioned earlier, this problem can be solved using the algorithms proposed in [29] or [198, 200].

One of the basic reasons why natural images can be compressed efficiently is the existence of high correlation among neighboring pixels. Therefore there is also a high correlation among neighboring mean values in the mean value QT decomposition. This dependency can be utilized by employing a first order DPCM between neighboring blocks in the QT. The DPCM is along the scanning path and hence if the scanning path is optimal as defined in section 4.4, the DPCM will be efficient. In chapter 5 we propose a motion estimation scheme where the inhomogeneous DVF is encoded using a QT. It is well known that the DVF has a high spatial correlation and hence we use a first order DPCM scheme to encoded it. The resulting motion estimation problem can then be formulated using the algorithm developed in this section.

4.5.2 Notation and assumptions

Let $q_{l,i} \in Q_{l,i}$ be the quantizer for block $b_{l,i}$, where $Q_{l,i}$ is the set of all admissible quantizers for block $b_{l,i}$. Let $s_{l,i} = [l, i, q_{l,i}] \in S_{i,l} = \{l\} \times \{i\} \times Q_{l,i}$ be the local state for block $b_{l,i}$, where $S_{l,i}$ is the set of all admissible state values for block $b_{l,i}$. Let $x = [l, i, q] \in X = \bigcup_{l=N}^{n_0} \bigcup_{i=0}^{4^{N-l}-1} S_{l,i}$ be the global state and X the set of all admissible state values. Finally let $x_0, \ldots, x_{N_T-1}$ be a global state sequence, which represents the left-to-right ordered leaves of a valid QT $\mathcal{T}$.

We assume that the frame distortion $D(x_0, \ldots, x_{N_{\mathcal{T}}-1})$ of the reconstructed frame is the sum of the individual block distortions $d(x_j)$, that is,

$$D(x_0, \ldots, x_{N_{\mathcal{T}}-1}) = \sum_{j=0}^{N_{\mathcal{T}}-1} d(x_j). \tag{4.31}$$

Most common distortion measures, such as the mean squared error (MSE), a weighted MSE or the peak signal to noise ratio (PSNR $= 10 * \log_{10}(\frac{255^2}{MSE})$) fall into this class.

As pointed out earlier, we allow for a first order dependency between the leafs along the scanning path. Because of the variable block sizes, we add the constraint that blocks which belong to the same parent, need to be scanned in sequence. This is the same constraint we used for deriving a procedure and a necessary and sufficient condition for an optimal scanning path in section 4.4. Therefore, the optimal scanning paths developed in section 4.4 are well suited for the algorithm proposed in this section. In particular, for all the examples presented, we employ an optimal scanning path which is based on a Hilbert curve as introduced in section 4.4.3.

Based on the first order DPCM assumption, the frame rate $R(x_0, \ldots, x_{N_{\mathcal{T}}-1})$ can be expressed as follows,

$$R(x_0, \ldots, x_{N_{\mathcal{T}}-1}) = \sum_{j=0}^{N_{\mathcal{T}}-1} r(x_{j-1}, x_j), \tag{4.32}$$

where $r(x_{j-1}, x_j)$ is the block rate which depends on the encoding of the current and previous blocks, since the QF of the current block depends on the quantizer selected for the previous block.

4.5.3 Optimal quad-tree decomposition for lossless schemes

In order to introduce the algorithm systematically, we first consider the problem when different lossless schemes are used to encode the blocks, instead of the quantizers. Since the reconstructed frame is identical to the original frame, the frame distortion is equal to zero and the goal is to minimize the required frame rate. This can be stated as follows,

$$\min_{x_0,\ldots,x_{N_T-1}} R(x_0,\ldots,x_{N_T-1}). \tag{4.33}$$

Since this is a lossless scheme, the blocks are not quantized, but encoded losslessly. With a slight abuse of notation, let $q_{l,i}$ represent different lossless encoding schemes for block $b_{l,i}$ (i.e., DPCM with different predictor order, etc.), instead of different quantizers. Since we will refer to this algorithm later on, we will still call the $q_{l,i}$'s quantizers in the following derivation.

Since we deal with a finite number of QTs and admissible quantizers, the above optimization problem can be solved by an exhaustive search. The time complexity for such an exhaustive search is extremely high. It can be shown that a lower bound for the number of different QTs is given by $2^{4^{N-n_0-1}}$, where $(N-n_0) \geq 1$. To find a lower bound on the total time complexity we consider the case when the QT is completely decomposed, i.e., the image is segmented into 4^{N-n_0} blocks of size $2^{n_0} \times 2^{n_0}$. We further assume that the cardinality $|S_{n_0,i}|$ of all admissible local states $S_{n_0,i}$ of the leafs is the same. Hence an exhaustive search, using only this single QT requires $|S_{n_0,i}|^{4^{N-n_0}}$ comparisons. Since we have only considered one QT, this is a lower bound on the time complexity for all QTs hence $\Omega_E(|S_{n_0,i}|^{4^{N-n_0}})$, where E stands for exhaustive search. We will show that the upper bound of the proposed algorithm is significantly smaller than this lower bound for an exhaustive search. Note that when we use the term time complexity, we refer to the number of comparisons necessary to find the optimal solution. This does not include the time spent to evaluate the operational rate distortion functions, since this strongly depends on the implementation.

4.5.4 Multilevel trellis

The form of the objective function (see Eq. (4.32)) of the optimization problem of Eq. (4.33) suggests that DP should be used to find the optimal solution efficiently. To be able to employ forward DP (the Viterbi algorithm), a DP recursion formula needs to be established. A graphical equivalent of the DP recursion formula (for first order neighborhoods, i.e., first order DPCM) is a trellis where the admissible nodes and the permissible transitions are explicitly indicated. Consider Fig. 4.18 which represents the multilevel trellis for a 32×32 image block ($N = 5$), with a QT segmentation developed down to level 3 ($n_0 = 3$, block size 8×8). The QT structure is indicated by the white boxes

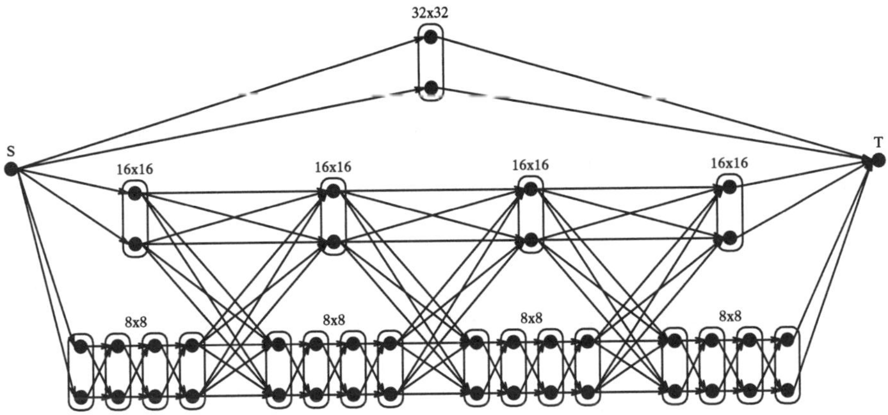

Figure 4.18 The multilevel trellis for $N = 5$ and $n_0 = 3$.

with the rounded corners. These white boxes are not part of the trellis used for the Viterbi algorithm but indicate the set of admissible state values $S_{l,i}$ for the individual blocks $b_{l,i}$. The black circles inside the white boxes are the nodes of the trellis (i.e., the state values $s_{l,i}$). Note that for simplicity, only two trellis nodes per QT node are indicated, but in general, a QT node can contain any number of trellis nodes. The auxiliary nodes, start and termination (S and T) are used to initialize the DPCM and to select the path with the smallest cost.

Each of the trellis nodes represents a different way of encoding the block it is associated with. Since the state of a block is defined to contain its level and number within that level (which identifies the blocks size and its position in the frame), and its quantizer, each of the nodes contains the rate (and distortion, for the lossy scheme) occurring encoding the associated block with the given quantizer.

As can be seen in Fig. 4.18, not every trellis node can be reached from every other trellis node. By restricting the permissible transitions, we are able to force the optimal path to select only valid QT decompositions. Such valid decompositions are based on the fact that at level l, block $b_{l,i}$ can replace four blocks at level $l-1$, namely $b_{l-1,4*i+0}, b_{l-1,4*i+1}, b_{l-1,4*i+2}$ and $b_{l-1,4*i+3}$. As we will see later in this section, the QT encoding cost can be distributed recursively over the QT so that each path picks up the right amount of QT segmentation overhead.

Assume that no QT segmentation is used and the block size is fixed at 8×8. In this case, only the lowest level in the trellis in Fig. 4.18 is used. The transition costs between the trellis nodes would be the rate required to encode the DPCM differences between consecutive blocks along the scanning path. Assume now that the next higher level, level 4, of the QT is included. Clearly the transition cost between the trellis nodes of level 3 stay the same. In addition, there are now transition costs between the trellis nodes of level 4 and also transition cost from trellis nodes of level 3 to trellis nodes of level 4 and vice versa, since each cluster of four blocks at level 3 can be replaced by a single block at level 4. The fact that a path can only leave and enter a certain QT level at particular nodes results in paths which all correspond to valid QT decompositions. Note that every QT node in a path is considered a leaf of the QT which is associated with this path.

In this example, a tree of depth 3 has been used to illustrate how the multilevel trellis is built. For QTs of greater depth, a recursive rule can be derived which leads to the proper connections between the QT levels. Since the Viterbi algorithm is a forward dynamic programming scheme, the optimal solution is found by going through the multilevel trellis from left to right. At any epoch of the Viterbi algorithm, the shortest path is found, over the set of all admissible nodes of the previous epoch (say $F_{l,i}$), to every node in the set of all admissible nodes at the current epoch (say $T_{l,i}$).

These two sets, at each epoch, the "from" set $F_{l,i}$ and the "to" set $T_{l,i}$ can be constructed recursively from the sets of the admissible state values $S_{l,i}$ for each block $b_{l,i}$. Consider Fig. 4.19 which shows a QT node and its four children. The $T_{l,i}$ set contains all the trellis nodes to which a transition from the left is permissible and $F_{l,i}$ contains all the trellis nodes from which a transition to the right is permissible. Clearly for the top QT level N, $T_{N,0} = S_{N,0}$ and $F_{N,0} = S_{N,0}$, which is the initialization of the recursion. As we will see in section 4.5.5, only the sets at level n_0 will be used in the Viterbi algorithm since they define the permissible transitions which in turn enforce the QT structure. These sets can be generated recursively in a Top-Down fashion by calling the function "sets(l, i)" below with the following parameters: sets$(N,0)$.

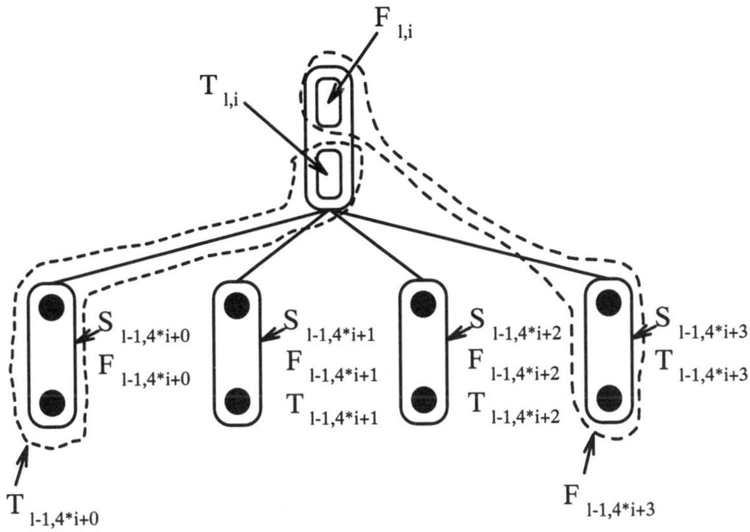

Figure 4.19 The recursive rule for the generation of the $T_{l,i}$ and $F_{l,i}$ sets.

sets(l, i)
  $T_{l-1,4*i+0} = T_{l,i} \cup S_{l-1,4*i+0}$;
  $F_{l-1,4*i+0} = S_{l-1,4*i+0}$;
  $T_{l-1,4*i+1} = S_{l-1,4*i+1}$;
  $F_{l-1,4*i+1} = S_{l-1,4*i+1}$;
  $T_{l-1,4*i+2} = S_{l-1,4*i+2}$;
  $F_{l-1,4*i+2} = S_{l-1,4*i+2}$;
  $T_{l-1,4*i+3} = S_{l-1,4*i+3}$;
  $F_{l-1,4*i+3} = F_{l,i} \cup S_{l-1,4*i+3}$;
  if $(l - 1 > n_0)$
    sets$(l - 1, 4 * i + 0)$;
    sets$(l - 1, 4 * i + 1)$;
    sets$(l - 1, 4 * i + 2)$;
    sets$(l - 1, 4 * i + 3)$;

In the presented multi level trellis, the nodes of the respective blocks hold the information about the rate (and in the case of a lossy scheme, the distortion) occurring when the associated block is encoded using the quantizer of the node. The rate needed to encode the DPCM encoding cost is incorporated into the transition cost between the nodes, but so far, the rate needed to encode the QT decomposition has not been addressed.

Since the Viterbi algorithm will be used to find the optimal QT decomposition, each node needs to contain a term which reflects the number of bits needed to split the QT at its level. Clearly, trellis nodes which belong to blocks of smaller size have a higher QT segmentation cost than nodes which belong to bigger blocks.

When the path includes only the top QT level N, then the QT is not split at all, and only one bit is needed to encode this. Therefore the segmentation cost $A_{N,0}$ equals one. For the general case, if a path splits a given block $b_{l,i}$ then a segmentation cost of $A_{l,i} + 4$ bits has to be added to its overall cost function, since by splitting block $b_{l,i}$, 4 bits will be needed to encode whether the four child nodes of block $b_{l,i}$ are split or not. Since the path only visits trellis nodes and not QT nodes, this cost has to be distributed to the trellis nodes of the child nodes of block $b_{l,i}$. How the cost is split among the child nodes is arbitrary since by the recursive definition of the $F_{l,i}$ and $T_{l,i}$ sets, every path which visits a sub-tree rooted by one child node, also has to visit the other three sub-trees rooted by the other child nodes. Therefore the path will pick up the segmentation cost, no matter how it has been distributed among the child nodes. Since the splitting of a node at level $n_0 + 1$ leads to four child nodes at level n_0, which can not be split further, no segmentation cost needs to be distributed among its child nodes. In other words, since it is known that the n_0 level blocks cannot be split, no information needs to be transmitted for this event.

These segmentation costs can be generated recursively in a Top-Down fashion by calling the function "seg(l, i)" below with the following parameters: seg(N,0). The recursion needs to be initialized with $A_{N,0} = 1$.

seg(l, i)
  $A_{l-1,4*i+0} = A_{l,i} + 4;$
  $A_{l-1,4*i+1} = 0;$
  $A_{l-1,4*i+2} = 0;$
  $A_{l-1,4*i+3} = 0;$
  if $(l - 1 > n_0)$
    seg$(l - 1, 4 * i + 0);$
    seg$(l - 1, 4 * i + 1);$
    seg$(l - 1, 4 * i + 2);$
    seg$(l - 1, 4 * i + 3);$

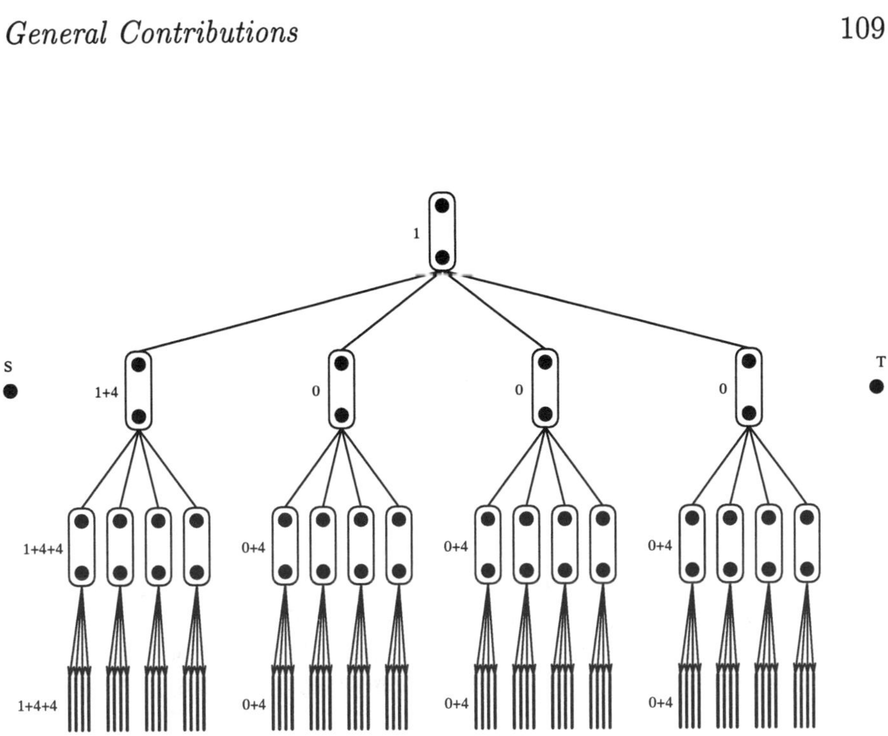

Figure 4.20 The recursive distribution of the quad-tree encoding cost among the trellis nodes for a quad-tree of depth 4.

Note that in the above function, the segmentation cost is distributed along the leftmost child. As mentioned before, any other assignment of the segmentation cost will lead to the same result. The recursion involved in the assignment of the encoding cost is illustrated in Fig. 4.20.

Having established the multi level trellis, the forward DP algorithm can be used to find the optimal state sequence $x_0^*, \ldots, x_{N_T-1}^*$ which will minimize the frame rate in problem (4.33).

4.5.5 Dynamic programming recursion formula

In the previous section, the multilevel trellis has been established which is a graphical equivalent of the DP recursion formula which is derived in this section. Again, each node x_j of the multilevel trellis contains $R^F(x_j)$, the rate needed to encode the quantized frame of the corresponding block $b_{l,i}$ using the quantizer $q_{l,i}$, which together with the level l and block number i defines the node. In the case of a lossy scheme, it also contains $D(x_j)$, which is the distortion occurring when the corresponding block is encoded using the given quantizer. As shown in the previous section, each node also contains $R^{SEG}(x_j) = A_{l,i}$, which is the recursively distributed QT encoding cost. In addition to that, the transition cost $R^{DPCM}(x_{j-1}, x_j)$ is the number of bits required for the encoding of the DPCM error between node x_{j-1} and node x_j, where $x_{j-1} \in F_{n_0,j-1}$ and $x_j \in T_{n_0,j}$. The dependency between the blocks is expressed by this term and it is the reason for using dynamic programming to solve the optimization problem. Note that $r(x_{j-1}, x_j)$ in Eq. (4.32) equals,

$$r(x_{j-1}, x_j) = R^{SEG}(x_j) + R^F(x_j) + R^{DPCM}(x_{j-1}, x_j). \tag{4.34}$$

As pointed out earlier, we will also use the algorithm presented in this section to solve the lossy scheme. Therefore let us introduce the generic functions $g(x_{j-1}, x_j)$ and $G(x_0, \ldots, x_{N_\mathcal{T}-1})$ which will be used to derive the DP recursion formula. For the lossless case, this generic functions are defined as follows,

$$g(x_{j-1}, x_j) = r(x_{j-1}, x_j) \tag{4.35}$$

and

$$G(x_0, \ldots, x_{N_\mathcal{T}-1}) = \sum_{j=0}^{N_\mathcal{T}-1} g(x_{j-1}, x_j). \tag{4.36}$$

The above discussion and these definitions imply that the following problem is identical to problem (4.33)

$$\min_{x_0, \ldots, x_{N_\mathcal{T}-1}} G(x_0, \ldots, x_{N_\mathcal{T}-1}). \tag{4.37}$$

The goal of DP is to find the optimal state sequence $x_0^*, \ldots, x_{N_\mathcal{T}-1}^*$ which will minimize Eq. (4.37). Let $g_k^*(x_k)$ be the minimum cost up to epoch k, where the epochs count the blocks $b_{n_0,k}$, hence $4^{N-n_0} - 1 \geq k \geq 0$,

$$g_k^*(x_k) = \min_{x_0, \ldots, x_{k-1}} \sum_{j=0}^{k} g(x_{j-1}, x_j), \tag{4.38}$$

where $x_{-1} = S$, the auxiliary starting node. The motivation behind this definition is the fact that $g_{4^{N-n_0}}^*(T)$ is the minimum cost required to traverse the trellis from (S) to (T), which is the solution to problem (4.37). From Eq. (4.38) follows that,

$$g^*_{k+1}(x_{k+1}) = \min_{x_0,\dots,x_k} \sum_{j=0}^{k+1} g(x_{j-1}, x_j) \tag{4.39}$$

$$= \min_{x_k} \left\{ \min_{x_0,\dots,x_{k-1}} \left(\sum_{j=0}^{k} g(x_{j-1}, x_j) + g(x_k, x_{k+1}) \right) \right\} \tag{4.40}$$

$$= \min_{x_k} \left\{ \min_{x_0,\dots,x_{k-1}} \left(\sum_{j=0}^{k} g(x_{j-1}, x_j) \right) + g(x_k, x_{k+1}) \right\} \tag{4.41}$$

which results in the DP recursion formula,

$$g^*_{k+1}(x_{k+1}) = \min_{x_k} \left(g^*_k(x_k) + g(x_k, x_{k+1}) \right). \tag{4.42}$$

Having established the DP recursion formula, the forward DP algorithm can be used to find the optimal state sequence. Note how the $F_{n_0,i}$ and $T_{n_0,i}$ sets are used to establish the connections between the layers of the multilevel trellis. The utilization of these sets constrains the admissible paths to represent a valid QT.

1. Initialization: for all $x_0 \in T_{n_0,0}$ do,

   $$g^*_0(x_0) = g(S, x_0), \tag{4.43}$$

   where S represents the starting node, i.e., an auxiliary node representing the convention for starting the DPCM. A pointer, $i(x_k)$ is introduced, which is used to keep track of the optimal path,

   $$i(x_0) = S. \tag{4.44}$$

2. Recursion: for $k = 0, \dots, 4^{N-n_0} - 2$ and for all $x_{k+1} \in T_{n_0,k+1}$ do,

   $$g^*_{k+1}(x_{k+1}) = \min_{x_k \in F_{n_0,k}} \left(g^*_k(x_k) + g(x_k, x_{k+1}) \right), \tag{4.45}$$

   $$i(x_{k+1}) = \arg \min_{x_k \in F_{n_0,k}} \left(g^*_k(x_k) + g(x_k, x_{k+1}) \right), \tag{4.46}$$

3. Termination:

$$g^*_{4^{N-n_0}}(T) = \min_{x_{4^{N-n_0}-1} \in F_{n_0,4^{N-n_0}-1}} (g^*_k(x_{4^{N-n_0}-1}) + g(x_{4^{N-n_0}-1}, T)), \tag{4.47}$$

where T is an auxiliary terminal node used to find the minimum of all optimal cost functions at epoch $4^{N-n_0} - 1$, as for our purpose, $g(x_{4^{N-n_0}-1}, T) = 0$. Again the back pointer is calculated

$$i(T) = \arg \min_{x_{4^{N-n_0}-1} \in F_{n_0,4^{N-n_0}-1}} (g^*_k(x_{4^{N-n_0}-1}) + g(x_{4^{N-n_0}-1}, T)). \tag{4.48}$$

4. Backtracking: the pointer i is used to backtrack the optimal state sequence.

$$k = N_T, \tag{4.49}$$

$$x^*_k = T, \tag{4.50}$$

do

$$k \leftarrow k - 1, \tag{4.51}$$

$$x^*_k = i(x^*_{k+1}), \tag{4.52}$$

until

$$x^*_k = S. \tag{4.53}$$

This leads to the optimal state sequence $x^*_0, \ldots, x^*_{N_T-1}$.

In Fig. 4.21, a QT of depth 4 is displayed and the optimal state sequence is indicated which leads to the segmentation shown in Fig. 4.22. Note that the resulting scanning path is optimal and the segmentation cost along the optimal path adds up to 13 bits, which is the number of bits needed to encode this QT decomposition. The bit stream for this QT decomposition is "1010000011001". The time complexity of DP is $O_{DP}(4^{N-n_0} * |S_{n_0,i}|^2)$ which is significantly smaller than the lower bound for the exhaustive search $\Omega_E(|S_{n_0,i}|^{4^{N-n_0}})$. Note that we have assumed that all the $S_{l,i}$ sets are of the same cardinality.

4.5.6 Optimal quad-tree decomposition for lossy schemes

So far we have considered lossless schemes. In this section we will apply the algorithms derived for a lossless scheme to find the optimal solution to a lossy

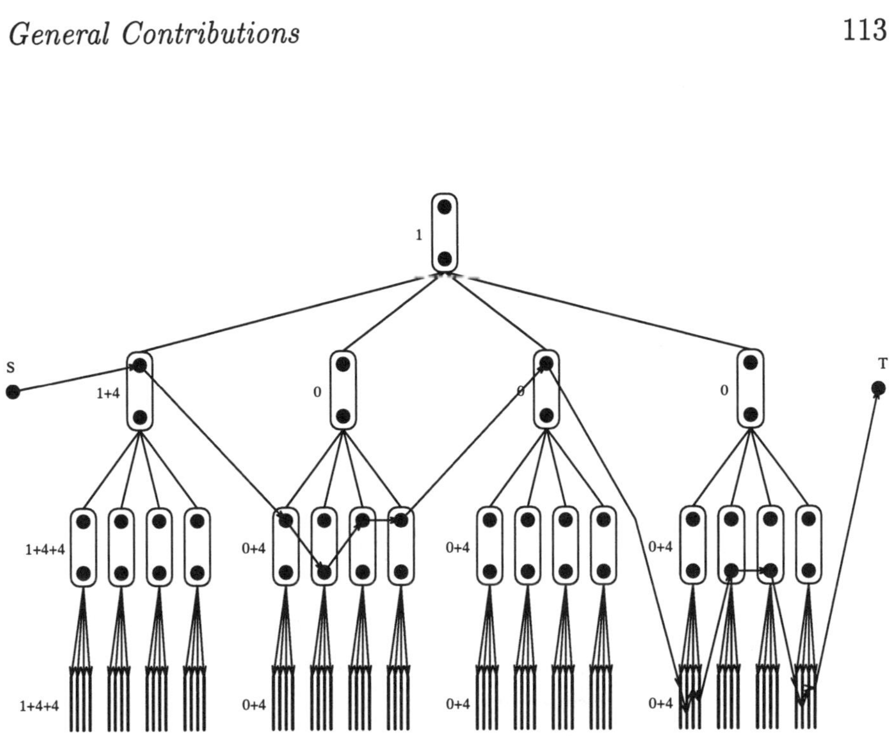

Figure 4.21 Optimal path through a quad-tree of depth 4.

scheme. Clearly for a lossy scheme it does not make sense to minimize the frame rate R with no additional constraints, since this would lead to a very high frame distortion D.

The most common approach to solve the tradeoff between the frame rate and the frame distortion is to minimize the frame distortion D subject to a given maximum frame rate R_{max}. This problem can be formulated in the following way,

$$\min_{x_0,\ldots,x_{N_\mathcal{T}-1}} D(x_0,\ldots,x_{N_\mathcal{T}-1}), \quad \text{subject to:} \quad R(x_0,\ldots,x_{N_\mathcal{T}-1}) \leq R_{max}. \tag{4.54}$$

In section 3.3 we introduced the Lagrangian multiplier method for discrete optimization problems, and we will now apply this method to find the convex hull solutions of the above problem. First we introduce the Lagrangian cost function which is of the following form,

$$J_\lambda(x_0,\ldots,x_{N_\mathcal{T}-1}) = D(x_0,\ldots,x_{N_\mathcal{T}-1}) + \lambda * R(x_0,\ldots,x_{N_\mathcal{T}-1}), \tag{4.55}$$

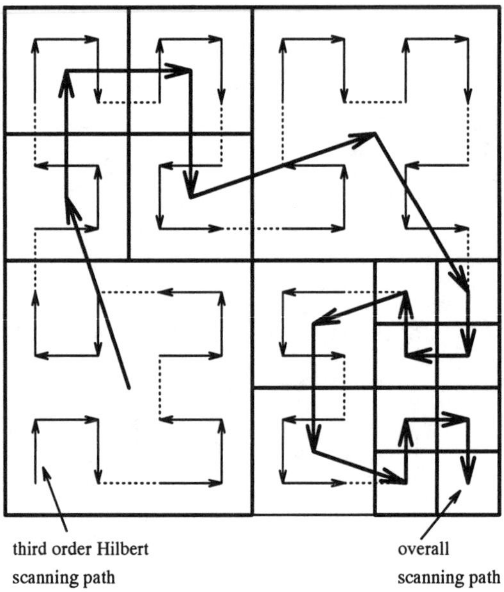

Figure 4.22 Quad-tree decomposition corresponding to the optimal state sequence.

where $\lambda \geq 0$ is called the Lagrangian multiplier. We have shown in section 3.3 that if there is a λ^* such that,

$$[x_0^*, \ldots, x_{N_{\mathcal{T}}-1}^*] = \arg \min_{x_0, \ldots, x_{N_{\mathcal{T}}-1}} J_{\lambda^*}(x_0, \ldots, x_{N_{\mathcal{T}}-1}), \tag{4.56}$$

and which leads to $R(x_0^*, \ldots, x_{N_{\mathcal{T}}-1}^*) = R_{max}$, then $x_0^*, \ldots, x_{N_{\mathcal{T}}-1}^*$ is also an optimal solution to the optimization problem stated in Eq. (4.54). It is well known that when λ sweeps from zero to infinity, the solution to problem (4.56) traces out the convex hull of the rate distortion curve, which is a non-increasing function. Hence bisection or the fast convex search we presented in section 4.2 can be used to find λ^*.

Therefore the problem at hand is to find the optimal solution to problem (4.56). We will show how the DP approach presented in the previous section can be modified to find the global minimum of problem (4.56). For a given λ, let the $g(x_{j-1}, x_j)$ function of Eq. (4.35) be defined as follows,

$$g(x_{j-1}, x_j) = d(x_j) + \lambda * r(x_{j-1}, x_j), \tag{4.57}$$

whereas in the lossless case,

$$r(x_{j-1}, x_j) = R^{SEG}(x_j) + R^F(x_j) + R^{DPCM}(x_{j-1}, x_j). \tag{4.58}$$

This implies that, $G(x_0, \ldots, x_{N_T-1}) = J_\lambda(x_0, \ldots, x_{N_T-1})$. Hence the DP algorithm presented in section 4.5.3 leads to the optimal solution of problem (4.56).

Note that the dual problem, which can be stated as follows,

$$\min_{x_0,\ldots,x_{N_T-1}} R(x_0, \ldots, x_{N_T-1}), \quad \text{subject to:} \quad D(x_0, \ldots, x_{N_T-1}) \leq D_{max}, \tag{4.59}$$

can be solved with exactly the same technique using the following relabeling of function names,

$$R(x_0, \ldots, x_{N_T-1}) \leftarrow D(x_0, \ldots, x_{N_T-1}), \tag{4.60}$$

$$D(x_0, \ldots, x_{N_T-1}) \leftarrow R(x_0, \ldots, x_{N_T-1}). \tag{4.61}$$

4.5.7 Example

In this section we use the mean value QT decomposition introduced earlier, as an example of how the above algorithm finds the optimal quantizers and the optimal QT decomposition simultaneously.

In the mean value QT decomposition, the frame is represented by differently sized blocks of a fixed luminance value. Usually it is implied that the luminance value is the quantized mean value of the original block. Conceptually, though, it could be any value, but clearly the distortion (we use the MSE) increases if another value than the mean is selected. Since these values are encoded using a DPCM scheme, where small prediction errors are encoded using short code words, it might be beneficial, in a rate distortion sense, to select a value slightly different than the mean value, since this might result in a large rate drop relative to the small distortion introduced.

In this example, we represent each luminance value as an integer between zero and 255. Therefore the set of admissible quantizers for each block is of size 256. On the other hand, if the value is not selected to be close to the actual mean, the resulting distortion is quite large, and hence the saving in rate cannot offset the increase in distortion. Let the mean of block $b_{l,i}$, rounded to the nearest integer, be denoted by $m_{l,i}$. We define the set of admissible quantizers for block $b_{l,i}$, $Q_{l,i} = \{m_{l,i} - 5, m_{l,i} - 4, \ldots, m_{l,i} + 5\}$. In other words, a block can be

Differential DC (absolute value)	Code word length Luminance	Code word length Chrominance
0	3	2
1	3	3
2 to 3	4	4
4 to 7	6	6
8 to 15	7	8
16 to 31	9	10
32 to 63	11	12
64 to 127	13	14
128 to 255	15	16

Table 4.2 Code word length for DC prediction error for the luminance and the chrominance channels.

Block size	PCM rate (bits)	DPCM rate (bits)	Distortion (PSNR)
8×8	3168	2560	23.46
4×4	12672	9080	26.28
2×2	50688	32573	30.20

Table 4.3 Comparison between mean value only encoding schemes for different, fixed block sizes, using the first luminance frame of the QCIF (176×144) "Mother and Daughter" sequence.

represented by its mean ± 5. Clearly for each of the different values a different distortion occurs.

The prediction error of the DPCM is encoded using the Huffman table suggested in [10] for the differential DC of the luminance. Table 4.2 shows how many bits are spent for a given absolute error in the luminance and the chrominance channels. In this example we are only interested in the luminance, but we will use the chrominance table in section 4.3.

Table 4.3 shows a comparison between mean value only encoding schemes, which use a fixed block size for the first luminance frame of the QCIF (176×144) "Mother and Daughter" sequence. In other words, the frame is segmented into 8×8, 4×4 or 2×2 blocks of which only the rounded mean values are transmitted.

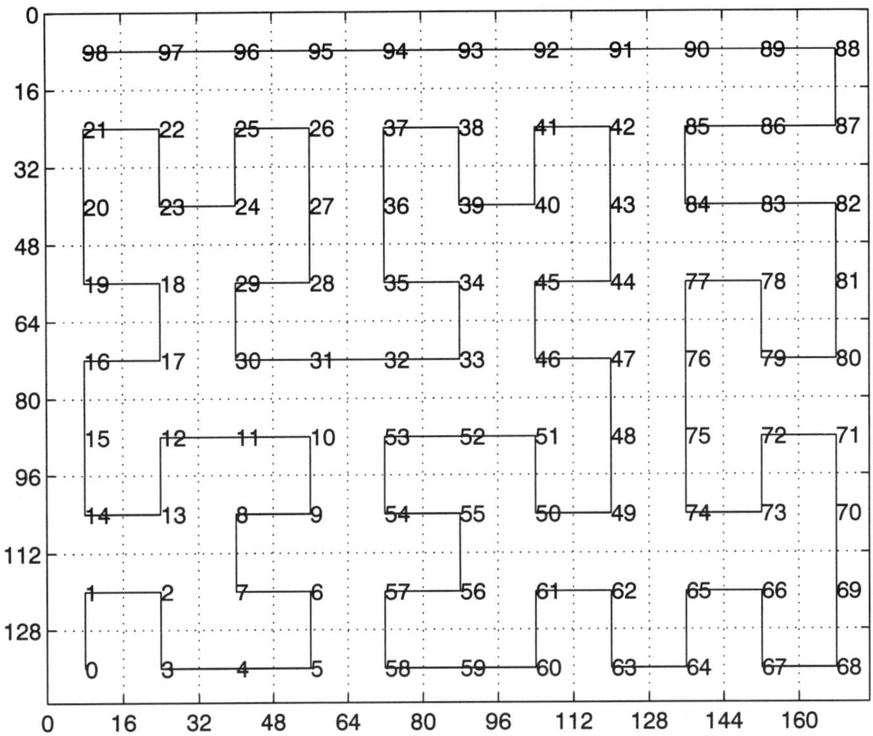

Figure 4.23 Modified Hilbert scan for a QCIF image, for blocks of size 16×16

In the "PCM rate" column, the resulting bit rate is stated when 8 bits per mean value are used. In the "DPCM rate" column, the above presented DPCM scheme is employed, which results in a smaller bit rate. Since a QCIF frame is not a square with a side length which is a power of two, a slightly modified Hilbert curve is used as scanning path. The scanning path for the 16×16 blocks, along which the first order DPCM is employed is displayed in Fig. 4.23. The scanning path for the smaller block sizes is recursively calculated using the orientations of the 16×16 blocks and the procedure described in section 4.4.3

The proposed optimal mean value QT decomposition with DPCM encoding of the luminance values finds the QT decomposition and the luminance values which result in the smallest distortion for a given bit rate. As we pointed out in section 3.3, the Lagrangian multiplier method can also be used to find the smallest bit rate for a given distortion. We apply this algorithm to the mean value QT decomposition, where the biggest block size is 16×16 and the smallest block size is 2×2. Therefore the original QCIF frame is represented by 99 QTs, since there are 99 different 16×16 blocks. In the first experiment, we set the distortion equal to the distortion of the 4×4 fixed block size implementation of Table 4.3, i.e., $D_{max} = 26.28$ dB PSNR. Using the optimal bit allocation algorithm, the distortion can be exactly matched and the required rate is 5049 bits, which is 44% fewer bits than the fixed block size implementation using the same DPCM encoding (9080 bits) and 60% fewer bits than the fixed block size implementation using pulse code modulation (PCM) encoding (12672 bits).

Now we compare this scheme with the discrete cosine transform (DCT) scheme used in H.263. For a quantizer step size of 10, H.263 results in a PSNR of 33.55 dB and a bit rate of 15008 bits for the first luminance frame of the QCIF "Mother and Daughter" sequence. Using the same mean value QT scheme as above, and fixing the rate to be equal to the H.263 rate, the resulting minimum distortion for this scheme is 29.52 dB PSNR, which is 4 dB lower than the DCT-based scheme. This was to be expected since we do not use a reconstruction filter as the ones employed in [195, 196, 210, 192], which can improve the reconstructed quality tremendously. The results of the H.263 scheme is displayed in Fig. 4.24 and in Fig. 4.25 the result of the optimal mean value QT encoding is displayed. The segmentation of the frame is shown in Fig. 4.26. As expected from a DCT-based scheme, one can observe the ringing artifacts in Fig. 4.24, whereas in Fig. 4.25 it is clear that the smallest block size of 2×2 is not fine enough for certain edges. But as pointed out above, the subjective quality of this scheme can easily be improved by using a reconstruction filter, which will reduces the blockiness.

4.5.8 The minimum maximum distortion criterion

So far we have presented a scheme which results in the optimal QT decomposition and quantizer selection in the case where the objective is to minimize the total distortion. We now introduce the adjustments necessary to the proposed approach, such that the resulting scheme is optimal for the minimum maximum distortion criterion. As we have shown in section 4.3, if we can solve the minimum rate problem optimally, then the minimum distortion problem can be solved optimally using an iterative approach. Hence we will only discuss the

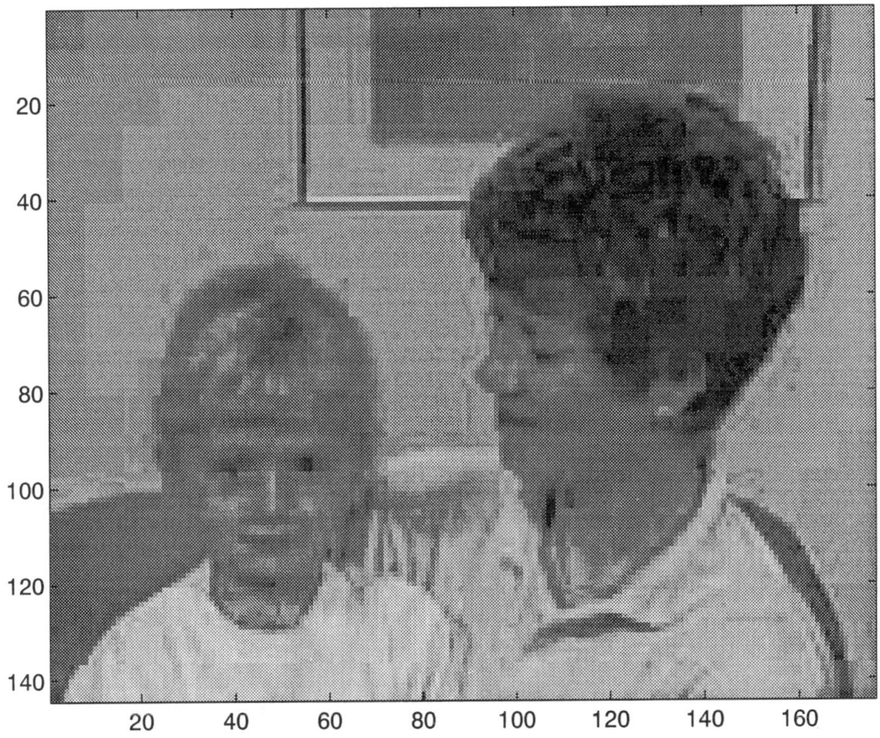

Figure 4.24 First luminance frame of the "Mother and Daughter" sequence encoded by H.263 using a quantizer step size of 10. The resulting bit rate is 15008 bits and the resulting distortion is 33.55 dB PSNR.

minimum rate case, which can be formulated as follows,

$$\min_{x_0,\ldots,x_{N_\mathcal{T}-1}} R(x_0,\ldots,x_{N_\mathcal{T}-1}), \qquad \text{subject to:} \quad D(x_0,\ldots,x_{N_\mathcal{T}-1}) \leq D_{max}, \tag{4.62}$$

where

$$D(x_0,\ldots,x_{N_\mathcal{T}-1}) = \max_{j\in[0,\ldots,N_\mathcal{T}-1]}\{d(x_j)\}. \tag{4.63}$$

Since the QT decomposition results in blocks of different sizes, it is important to realize how the block distortions $d(x_j)$ should be defined so that the desired effect of a best "worst case" is achieved. We propose to define the block distortion $d_{l,i}(q)$ (recall that $x = [l, i, q]$, hence $d(x_j)$ is equivalent to $d_{l,i}(q)$) of block

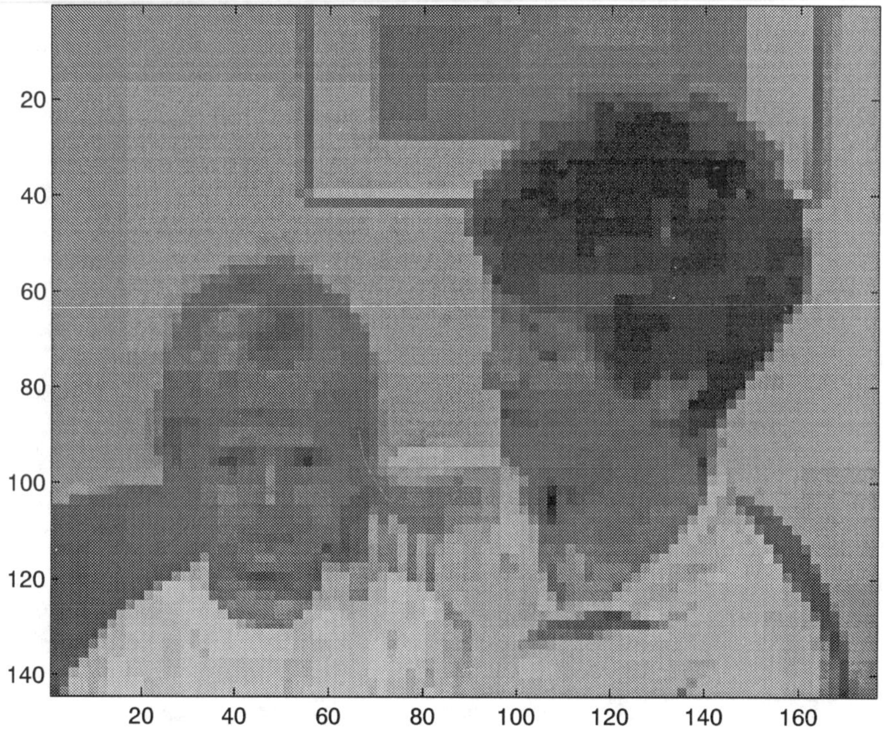

Figure 4.25 First luminance frame of the "Mother and Daughter" sequence encoded by the optimal mean value QT decomposition. The resulting bit rate is 14996 bits and the resulting distortion is 29.52 dB PSNR.

$b_{l,i}$ which is encode using the quantizer q, in the following recursive fashion, where $d_{n_0,i}(q)$ is the MSE (or signal to noise ratio, SNR) of block $b_{n_0,i}$,

$$d_{l,i}(q) = \max_{k \in [0,\ldots,3]} \{d_{l-1,4*i+k}(q)\}. \tag{4.64}$$

Therefore the MSE of every block at level n_0 is guaranteed to be below or equal to D_{max}, and this is achieved with the fewest bits possible.

The above algorithm can be changed to find the optimal solution for the minimum rate case and the minimum maximum distortion criterion by redefining the $g(x_{j-1}, x_j)$ function of Eq. (4.35). For a given D_{max}, let the $g(x_{j-1}, x_j)$

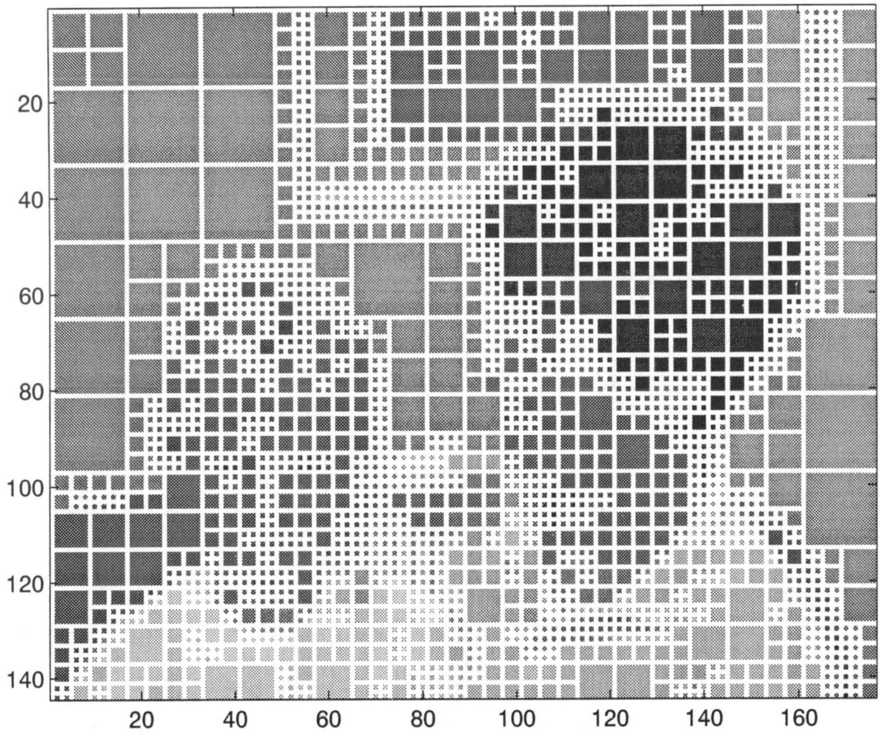

Figure 4.26 Segmentation of the first luminance frame of the "Mother and Daughter" sequence encoded by the optimal mean value QT decomposition.

function be defined as follows,

$$g_i(x_{j-1}, x_j) = \begin{cases} \infty & : \quad d(x_j) > D_{max} \\ r(x_{j-1}, x_j) & : \quad d(x_j) \leq D_{max} \end{cases}, \tag{4.65}$$

where as before $r(x_{j-1}, x_j)$ equals $R^{SEG}(x_j) + R^F(x_j) + R^{DPCM}(x_{j-1}, x_j)$. This implies that, $G(x_0, \ldots, x_{N_\mathcal{T}-1})$ of Eq. (4.36) is equal to the number of bits needed to encode the frame, given that there exists at least one quantizer sequence such that each block distortion of level n_0 is smaller or equal to the maximum permissible distortion D_{max}. We have shown that the proposed algorithm finds a QT decomposition and a quantizer selection such that $G(x_0, \ldots, x_{N_\mathcal{T}-1})$ is minimized. Hence with $G(x_0, \ldots, x_{N_\mathcal{T}-1})$ being the bit rate for a given maximum distortion, the above redefinition of $g_i(x_{j-1}, x_j)$ results in an algorithm which finds the smallest bit rate for a given maximum distortion using the minimum maximum distortion criterion.

As pointed out above and discussed in section 4.3, the minimum distorting problem for the minimum maximum distortion criterion can now be solved by iteratively solving the minimum rate problem as discussed in this section.

4.6 SUMMARY

In this chapter we presented the general contributions of this monograph. First we generalized the theory of optimal bit allocation for dependent quantizers to include dependencies of arbitrary order. We then introduced a very fast convex search for the optimal tradeoff parameter in Lagrangian optimization problems. The proposed search is based on a second order Bezier curve and exhibits very fast convergence. We proposed an optimal bit allocation scheme for dependent quantizers, where the objective is to minimize the maximum distortion. We introduced the notion of an optimal scanning path and we derived a procedure and a necessary and sufficient condition to generate an optimal scanning path for an arbitrary QT decomposition. We then introduced a special scanning path for a QT which is based on a Hilbert curve. This scanning path is optimal and the resulting blocks are square. We proposed a scheme for the optimal decomposition of a QT in the rate distortion sense, for the case when there are dependencies between the QT leafs. All of the above schemes will be used in the remainder of this monograph to develop solution approaches for specific problems, although, as we pointed out in this chapter, theses schemes can have a wide range of application.

5

OPTIMAL MOTION ESTIMATION AND MOTION COMPENSATED INTERPOLATION FOR VIDEO COMPRESSION

In this chapter we introduce two motion estimation schemes which are optimal in the rate distortion sense. The first is an optimal region matching scheme, and the second is an optimal quad-tree (QT) based motion estimator. We also introduce a motion compensated interpolation scheme, which employs the QT-based motion estimator. The interpolation scheme offers full control of the tradeoff between a global smoothness constraint and the interpolation error energy.

We use operational rate distortion theory to formulate the motion estimation problem for video coding, since there is an inherent tradeoff between the displaced frame difference (DFD) energy (distortion) and the displacement vector field (DVF) bit rate. Clearly the more bits we use to describe the motion in the scene, the smaller the DFD energy is.

As we pointed out in section 2.2.1, the motion estimation problem for video coding is quite different from the general motion estimation problem. There are several reasons for this. First, for video coding, the performance of a motion estimation scheme can be assessed using a distortion measure, whereas for the general motion estimation problem, such measures are very hard to find. Second, for video coding, the DVF has to be transmitted and therefore the minimization of the distortion is constrained by the available number of bits to encode the DVF, whereas for the general motion estimation problem, the constraints are used to enforce certain desirable properties of the DVF, such as smoothness. Third, for video coding, the resulting DVF can be arbitrary, i.e., it does not have to correspond to the real motion in the scene, as long as the distortion is minimized for the available bit rate. The goal of the general motion estimation problem is in strong contrast to this, since it is to find the two dimensional projection onto the image plane of the real world three dimensional motion. Since the three dimensional motion is usually unknown, the performance of the scheme needs to be evaluated by inspecting the resulting motion field and making a subjective judgment about its quality. Another evaluation scheme is based on motion compensated interpolation, where the estimated DVF is used to interpolate a known frame to which the interpolated frame is then compared. Clearly the better the DVF describes the real motion, the closer the interpolated frame matches the known frame.

This chapter is organized as follows: In section 5.1 we introduce a region matching scheme which is optimal in the rate distortion sense. In section 5.2 we extend the region-based scheme to a QT-based motion estimator, which is also optimal in the rate distortion sense. In section 5.3 we use the QT-based motion estimator to formulate a motion compensated interpolation scheme and in section 5.4 we summarize this chapter.

5.1 OPTIMAL REGION MATCHING

In this section we introduce a region matching scheme which, for a given maximum rate R_{max} to encode the DVF, finds the DVF resulting in the smallest DFD energy. The motion estimation algorithm is based on the optimal bit allocation scheme for dependent quantizers we introduced in section 4.1. We focus the discussion on the minimum total distortion objective, since this corresponds to minimizing the total energy of the DFD. Nevertheless, the minimum maximum distortion scheme for dependent quantizers, proposed in section 4.1 can also be used in the presented framework.

In [67], the problem of rate-constrained motion estimation is considered and the optimal bit allocation condition for a strictly convex and everywhere differentiable multivariate rate distortion function is derived. It is applied to the problem of optimal bit allocation between the DVF and the DFD and a rate-constrained, region-based motion estimator is introduced. The derivation in [67] is true to the idea of rate distortion theory, where certain models are assumed and then solutions are derived based on these models.

In this monograph, we follow the paradigm of operational rate distortion theory, where an efficient scheme is derived and its decisions are analyzed such that the resulting overall approach is optimal. Therefore, we do not assume knowledge of a convex and everywhere differentiable multivariate rate distortion function, but instead we deal with finite sets of admissible motion vectors.

5.1.1 Notation and assumptions

We assume that the current frame is segmented into N regions, $o_0, \ldots, o_{N-1}$, and that this segmentation and the associated scanning path are known to both the encoder and the decoder. Note that these regions do not need to be blocks, but most commonly that is what they are. We then number the regions such that the scanning path visits them in ascending order. Every region o_i has a motion vector $m_i \in M_i$ associated with it, where M_i is the set of all admissible motion vectors for region o_i. As in every practical video coding scheme, we assume that the set M_i is finite.

The DFD energy of frame k, $D_k(m_0, \ldots, m_{N-1})$, in a motion compensated video coder, is a function of the previously reconstructed frame $\hat{f}_{k-1}$, the DVF $= \{m_i\}_{i=0}^{N-1}$, and the current frame f_k. Equivalently, the DVF rate $R_k(m_0, \ldots, m_{N-1})$ is also a function of $\hat{f}_{k-1}$, f_k and the DVF.

The next assumption expresses the idea that the frame DVF rate and frame DFD energy can be decomposed into a sum of region rates $r_i(m_{i-a}, \ldots, m_{i+b})$ and region energies $d_i(m_{i-a}, \ldots, m_{i+b})$, which only depend on a local neighborhood. We assume that there exist integers $a \geq 0$, and $b \geq 0$ and a family of functions d_i and r_i such that the following holds,

$$D_k(m_0, \ldots, m_{N-1}) = \sum_{i=0}^{N-1} d_i(m_{i-a}, \ldots, m_{i+b}), \tag{5.1}$$

$$R_k(m_0, \ldots, m_{N-1}) = \sum_{i=0}^{N-1} r_i(m_{i-a}, \ldots, m_{i+b}), \tag{5.2}$$

where the motion vectors m_j not belonging to any region ($j \notin [0, \ldots, N-1]$), represent the boundary conditions of the above sums and can be set to any desired value. The above assumption is very important since the efficiency of the optimization procedure directly depends on the value $(a+b+1)$ which defines the size of the neighborhood.

It is noted here that assumptions (5.1) and (5.2) are quite general and valid for every existing video coding standard. For example, in MPEG-1 and MPEG-2, the DVF rate for a given block depends not only on the motion vector of that block, but also on the motion vector of the previous block, which is used as a predictor for the current motion vector.

Usually the region energy measured by the popular mean squared error (MSE) measure depends only on the motion vector of the current block. A noteworthy exception is H.263 when the "Advanced Prediction Mode" is used. Then, overlapped block motion compensation is employed and the MSE of a given 8×8 block depends now on four spatial neighbors.

5.1.2 Optimal bit allocation

We propose to solve the tradeoff between the DVF rate and the DFD energy by minimize the DFD energy D_k subject to a given maximum DVF rate R_{max}. This problem can be formulated in the following way,

$$\min_{m_0,\ldots,m_{N-1}} D_k(m_0,\ldots,m_{N-1}), \qquad \text{subject to:} \quad R_k(m_0,\ldots,m_{N-1}) \leq R_{max}. \tag{5.3}$$

By selecting a particular motion vector m_i for region o_i, this region is represented in the predicted frame by a region of the same size, but at a different location, in the previously reconstructed frame. This prediction of region o_i can be considered as a quantization of region o_i, where a certain rate is used and a certain distortion occurs. If the MSE is used as the distortion measure, then the distortion is equivalent to the DFD energy. Hence the rate constrained motion estimation problem can be viewed as an optimal bit allocation problem among dependent quantizers. Therefore the scheme introduced in section 4.1 can be directly applied to find the optimal DVF.

The optimal bit allocation scheme for the minimum total distortion presented in section 4.1 is based on the Lagrangian multiplier method for discrete constrained optimization problems reviewed in section 3.3. The basic idea of Lagrangian relaxation is to move the "hard" constraint, weighted by a Lagrangian multiplier λ, into the objective function. This results in the relaxed problem,

$$\min_{m_0,\ldots,m_{N-1}} \left(D_k(m_0,\ldots,m_{N-1}) + \lambda * R_k(m_0,\ldots,m_{N-1})\right), \tag{5.4}$$

which is an unconstrained optimization problem. When the relaxed problem is much easier to solve than the constrained problem, then the Lagrangian multiplier method is a powerful tool. Since we assume that the frame rate and the frame distortions are the sums of the region rates and the region distortions, which in turn only depend on their local neighborhoods, the unconstrained problem can be solved optimally and efficiently using dynamic programming (see section 3.4). Being able to solve the relaxed problem optimally using dynamic programming, the optimal solution (or its convex hull approximation) to the constrained problem can be found by an iterative search, for example the very fast convex search we presented in section 4.2.

Note that the relaxed problem is very similar to the regularization approach often used in motion estimation. Since motion estimation is an ill posed problem, it is common to regularize it by adding additional terms to the objective function, which express prior knowledge of the solution. In fact, it is quite common to minimize the DFD energy plus a term which reflects the smoothness of the DVF, weighted by a regularization parameter λ. This is exactly the form of the relaxed problem and we will use this fact in section 5.3 to formulate a motion compensated interpolation scheme. Recall that for the proposed rate constrained motion estimator, the relaxed problem is solved iteratively for different λs until the rate constraint can be satisfied. For this iteration to converge, it is very important that the relaxed problem is solved optimally. In motion estimation, the regularization parameter is usually fixed and only a local minimum can be found, except when the objective function is convex and a continuous framework is used.

As for all Lagrangian approaches, the dual problem, which can be stated as follows,

$$\min_{m_0,\ldots,m_{N-1}} R_k(m_0,\ldots,m_{N-1}), \quad \text{subject to:} \quad D_k(m_0,\ldots,m_{N-1}) \leq D_{max}, \tag{5.5}$$

can be solved with exactly the same technique using the following relabeling of function names,

$$R_k(m_0,\ldots,m_{N-1}) \leftarrow D_k(m_0,\ldots,m_{N-1}), \tag{5.6}$$

$$D_k(m_0,\ldots,m_{N-1}) \leftarrow R_k(m_0,\ldots,m_{N-1}). \tag{5.7}$$

5.1.3 Example

We present a block-based (16 × 16) example of an optimal motion estimation scheme based on the above proposed scheme. We use a first order differential pulse code modulation (DPCM) to encode the DVF. In other words, the current motion vector is predicted using the previous one. The only exception is the encoding of the zero motion vectors, for which a special bit is reserved, since this is the most common event. The bit is set to one if the motion vector is zero and to zero otherwise. This first order DPCM results in the following expression for the DVF rate,

$$R_k(m_0,\ldots,m_{N-1}) = \sum_{i=0}^{N-1} r_i(m_{i-1},m_i). \tag{5.8}$$

Since the distortion measures the energy in the DFD (MSE), it can be expressed as follows,

$$D_k(m_0,\ldots,m_{N-1}) = \sum_{i=0}^{N-1} d_i(m_i), \tag{5.9}$$

where the block distortions $d_i(m_i)$, are the MSE occurring when the motion vector m_i is used.

This DVF encoding scheme is similar to the one used in test model 4 (TMN4 [47]) for the H.263 standard. The main differences are that in this example, the scanning path is not a raster scan as in H.263 but the modified Hilbert scan displayed in Fig. 4.23 for a quarter common intermediate format (QCIF) sequence and that only a first order DPCM encoding is used and not a vector median-based one. The Hilbert scan results in consecutive motion vectors with higher correlation and hence the DPCM scheme is more efficient. The first order DPCM results in a faster optimization procedure since the neighborhood is as small as possible for a predictive scheme (see section 4.1, a=1, b=0). The motion estimation has an accuracy of 0.5 pixels, where the half pixel values are calculated using bilinear interpolation. The maximum range of the motion vectors is ± 15.5 pixels.

For the presented experiment, we use the 176-th and 180-th frames of the QCIF sequence "Mother and Daughter" which are displayed in Fig. 5.1. We only use the luminance part of the sequence to perform the experiments, but the presented theory also covers the case when the chrominance distortion is include in the distortion measure. It therefore provides an elegant approach for fusing the information provided by the different channels.

The region distortion $d_i(m_{i-a},\ldots,m_{i+b})$ can be defined arbitrarily and hence it can contain contributions from the luminance channel as well as from the two chrominance channels. In the most general form, the region distortion can be written as follows,

$$d_i(m_{i-a},\ldots,m_{i+b}) = f(d_i^Y(m_{i-a},\ldots,m_{i+b}), d_i^{Cb}(m_{i-a},\ldots,m_{i+b}), d_i^{Cr}(m_{i-a},\ldots,m_{i+b})) \tag{5.10}$$

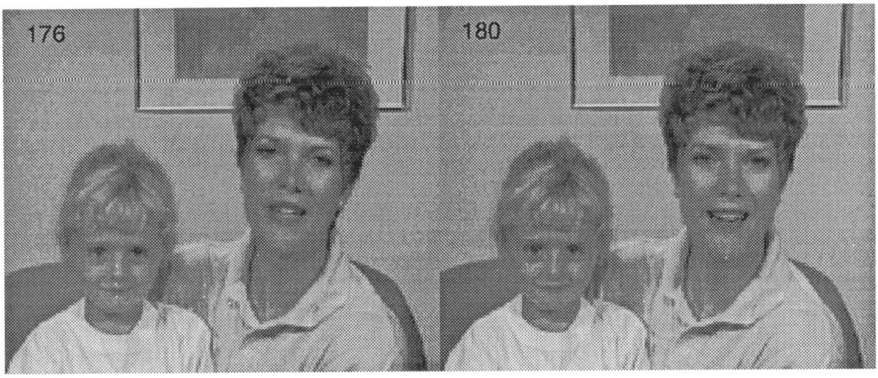

Figure 5.1 The original 176-th and 180-th frames of the QCIF sequence "Mother and Daughter".

where $f(\cdot)$ is an arbitrary function and the superscripts Y, Cb and Cr indicate the distortions in the luminance and chrominance channels. One popular choice for the function $f(\cdot)$ is a weighted sum,

$$\begin{aligned} d_i(m_{i-a}, \ldots, m_{i+b}) &= d_i^Y(m_{i-a}, \ldots, m_{i+b}) + \alpha d_i^{Cb}(m_{i-a}, \ldots, m_{i+b}) \\ &\quad + \beta d_i^{Cr}(m_{i-a}, \ldots, m_{i+b}), \end{aligned} \tag{5.11}$$

since for this definition of the region distortion, the frame distortion is also the weighted sum of the luminance and chrominance frame distortions. Clearly the selection of an appropriate α and β has to be done experimentally. Note that by using a region distortion as defined above, the DVF is found optimally, using the information from all three channels. Nevertheless, in the presented results, we set α and β to zero.

We intra code the 176-frame using TMN4 and a quantizer step size of 10. The resulting peak signal to noise ratio (PSNR) for this frame is 33.85 dB. Then the original 180-th frame is used to find the DVF. First we employ the TMN4 block matching scheme which uses the sum of the absolute error and favors the zero motion vector by reducing its error by a constant amount of 100. The resulting bit rate for the DVF and the DFD energy are listed in row "TMN4 (16×16)" in Table 5.1. The predicted frame and the DVF are displayed in Fig. 5.2.

	Rate (bits)	Dist. (PSNR)
TMN4 (16 × 16)	470	30.65
matched Rate (16 × 16)	470	30.80
matched Dist.(16 × 16)	377	30.65
matched Rate (QT)	472	31.31
matched Dist (QT)	344	30.72

Table 5.1 Comparison between the TMN4 motion estimation algorithm and the proposed optimal motion estimators using the luminance values of the 176-th and 180-th frames of the QCIF sequence "Mother and Daughter".

From a theoretical point of view, all possible motion vectors should belong to the set of admissible motion vectors for a given block. With a search window of ±15.5 pixels and an accuracy of 0.5 pixels, there are 3969 admissible motion vectors per block. Since the time complexity is quadratic in the number of admissible motion vectors, (see section 4.1, a=1, b=0) the fewer motion vectors we use, the faster the solution is found. Clearly, most motion vectors are not likely candidates for the optimal motion vector since they do not represent the real motion in the scene and hence result in a high rate and a high distortion. We therefore propose the following selection procedure for the set of admissible motion vectors. First, we find the K motion vectors with integer accuracy which result in the smallest MSEs and then the set of all admissible motion vectors M_i of block i is defined as these K motion vectors and all their half-pixel neighbors. By setting $K = 10$, the performance degradation is negligible, but the algorithm is much faster than when all the admissible motion vectors are used. Note that this use of prior knowledge about the optimal solution can be included in the optimization procedure, since dynamic programming can handle such hard constraint sets easily, whereas for a variational scheme, such constraints are very hard to incorporate.

Now the optimal block-based (16 × 16) motion estimator is run, where first the maximum rate is set equal to the TMN4 rate, i.e., $R_{max} = 470$ bits. We call this experiment "matched Rate (16 × 16)" in Table 5.1 and the resulting rate and distortion are listed in the corresponding row. Note that for the same rate, the total distortion is reduced, or in other words, a better prediction is achieved. This is also apparent from the resulting predicted frame and the DVF which are displayed in Fig. 5.3.

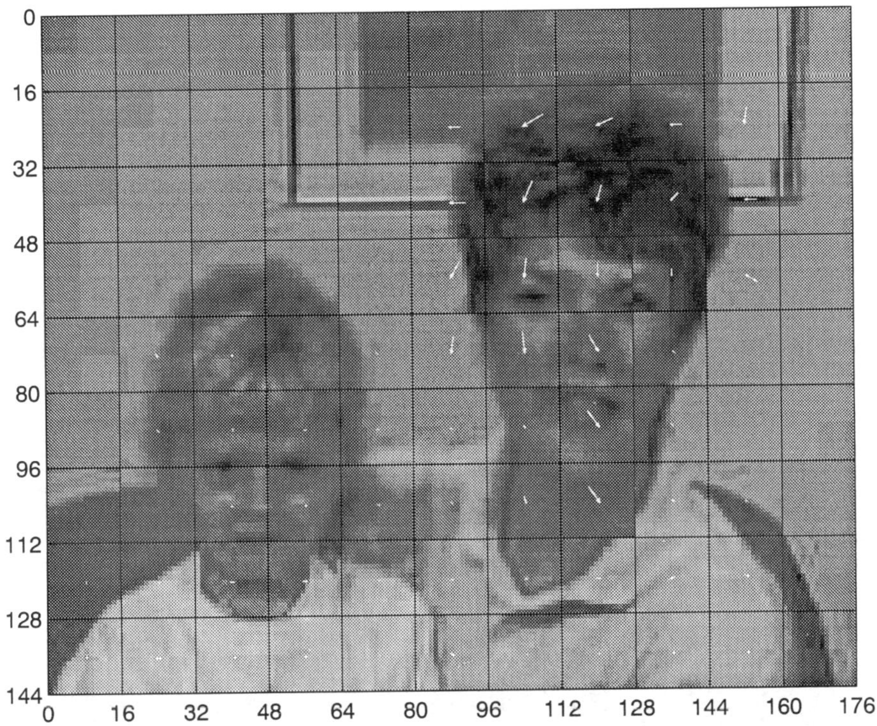

Figure 5.2 The predicted frame and the DVF for TMN4 block matching.

For the next experiment, the DFD energy (the distortion) of the TMN4 run is matched by setting $D_{max} = 30.65$ dB PSNR. We call this experiment "matched Dist. (16×16)" in Table 5.1 and the resulting rate and distortion are listed in the corresponding row. Note that for the same distortion, the bit rate is reduced significantly. The resulting predicted frame and the DVF are displayed in Fig. 5.4. Even though the DFD energy is the same as in Fig. 5.2, the predicted frame is of higher quality. This can be explained by observing that the entropy code used for the motion vector differences results in short code words for small differences and long code words for big differences. Therefore a smooth DVF is always favored since it can be encoded more easily. A discontinuity is only allowed if in return the distortion drops significantly. Also the DVF is encoded along the modified Hilbert scan and hence the smoothness of the DVF is not only in one direction, but it is a global smoothness.

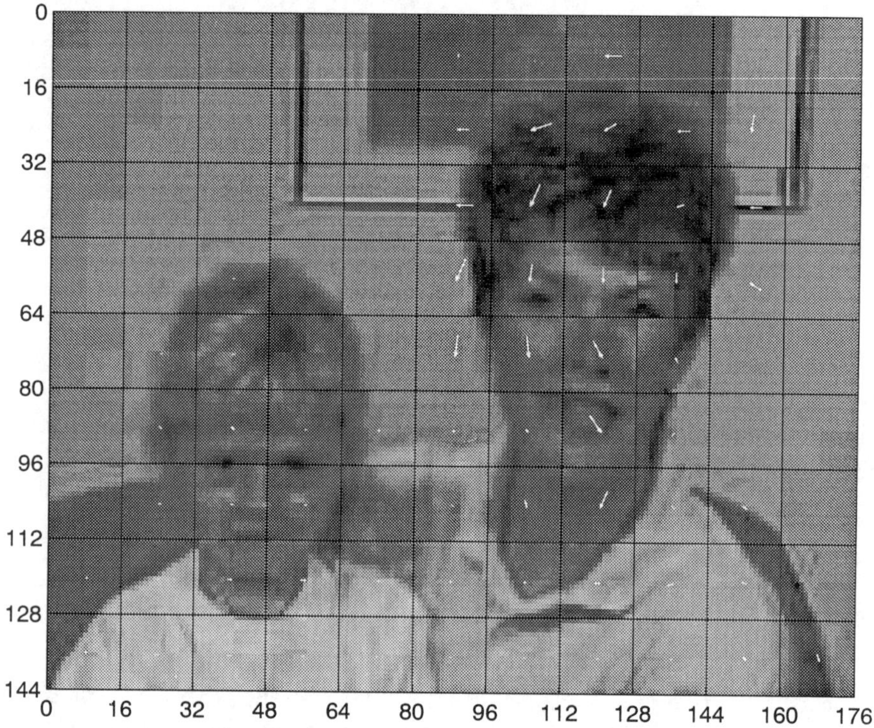

Figure 5.3 The predicted frame and the DVF for the optimal block matching algorithm, when the rate matches the TMN4 rate.

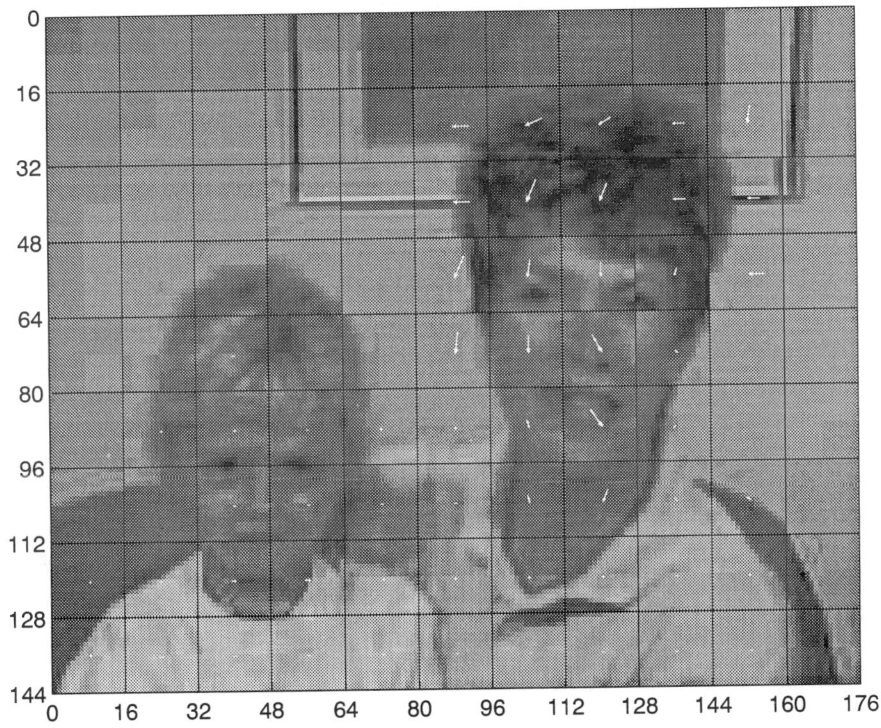

Figure 5.4 The predicted frame and the DVF for the optimal block matching algorithm, when the distortion matches the TMN4 distortion.

5.2 OPTIMAL QT-BASED MOTION ESTIMATOR

In this section we extend the scheme presented in the previous section to a variable block-based motion estimator, where the variable block sizes are encoded using a QT. We use the optimal bit allocation algorithm for QTs with leaf dependencies presented in 4.5 to find the optimal motion vectors and the optimal QT decomposition. In section 2.3, we argued that an efficient representation of a DVF has to be inhomogeneous, since in areas with uniform motion a single motion vector is an adequate description of the motion in the scene, while in areas with complex motion, a dense DVF is required. Clearly a QT-based variable block size motion estimator results in an inhomogeneous representation of the DVF and hence it should lead to a more compact description of the motion in the scene.

5.2.1 Notation and assumptions

We assume that the current frame is segmented by a QT and we use the basic QT notation introduced in section 4.4.1. We also introduce additional notation which follows closely the one used in section 4.5.2. Let $m_{l,i} \in M_{l,i}$ be the motion vector for block $b_{l,i}$, where $M_{l,i}$ is the set of all admissible motion vectors for block $b_{l,i}$. Let $s_{l,i} = [l, i, m_{l,i}] \in S_{i,l} = \{l\} \times \{i\} \times M_{l,i}$ be the local state for block $b_{l,i}$, where $S_{l,i}$ is the set of all admissible state values for block $b_{l,i}$. Let $x = [l, i, q] \in X = \bigcup_{l=N}^{n_0} \bigcup_{i=0}^{4^{N-l}-1} S_{l,i}$ be the global state and X the set of all admissible state values. Finally let $x_0, \ldots, x_{N_\mathcal{T}-1}$ be a global state sequence, which represents the left-to-right ordered leaves of a valid QT $\mathcal{T}$.

We assume that the frame distortion $D(x_0, \ldots, x_{N_\mathcal{T}-1})$ is the sum of the individual block distortions $d(x_j)$, that is,

$$D(x_0, \ldots, x_{N_\mathcal{T}-1}) = \sum_{j=0}^{N_\mathcal{T}-1} d(x_j). \tag{5.12}$$

Again, the block distortions measures the MSE error in the DFD and hence the frame distortion is the energy of the DFD.

Since we will encode the DVF using a first order DPCM we allow for a first order dependency between the leafs along the scanning path. Because of the variable block sizes, we add the constraint that blocks which belong to the same parent, need to be scanned in sequence. This is the same constraint we used for deriving a procedure and a necessary and sufficient condition for an optimal scanning path in section 4.4. Therefore, the optimal scanning paths developed in section 4.4 are well suited for the algorithm proposed in this section. In particular, for all the examples presented, we employ an optimal scanning path which is based on a Hilbert curve as introduced in section 4.4.3.

Based on the first order DPCM assumption, the frame rate $R(x_0, \ldots, x_{N_{\mathcal{T}}-1})$ can be expressed as follows,

$$R(x_0, \ldots, x_{N_{\mathcal{T}}-1}) = \sum_{j=0}^{N_{\mathcal{T}}-1} r(x_{j-1}, x_j), \tag{5.13}$$

where $r(x_{j-1}, x_j)$ is the block rate which depends on the encoding of the current and previous blocks, since the quantizer function of the current block depends on the motion vector selected for the previous block.

Note that in section 5.1, the rate and distortion where allowed to have any order of dependency. In this extension, we have to restrict the dependencies to first order so that the scheme introduced in section 4.5 can be applied. Again, the block distortion measures the MSE error in the DFD and hence the frame distortion is the energy of the DFD. By considering every possible motion vector of a given block as an admissible quantizer, the optimal bit allocation scheme for QTs with leaf dependencies can now be used to find the optimal inhomogeneous DVF, where the DVF block size is encoded using a QT.

5.2.2 Optimal bit allocation

As in section 5.1, we state the motion estimation problem for video coding as the following constrained optimization problem,

$$\min_{x_0, \ldots, x_{N_{\mathcal{T}}-1}} D(x_0, \ldots, x_{N_{\mathcal{T}}-1}), \quad \text{subject to:} \quad R(x_0, \ldots, x_{N_{\mathcal{T}}-1}) \leq R_{max}. \tag{5.14}$$

By selecting a particular motion vector $m_{l,i}$ for block $b_{l,i}$, this block is represented in the predicted frame by a block of the same size, but at a different location, in the previously reconstructed frame. This prediction of block $b_{l,i}$ can be considered as a quantization of block $b_{l,i}$, where a certain rate is used and a certain distortion occurs. If the MSE is used as the distortion measure, then the distortion is equivalent to the DFD energy. Hence the rate constrained motion estimation problem can be viewed as an optimal bit allocation problem among the blocks (leaves) of a QT with leaf dependencies. Therefore the scheme introduced in section 4.5 can be directly applied to find the optimal inhomogeneous DVF.

Since this is a Lagrangian relaxation based scheme, the proposed approach can also solve the dual problem, which can be stated as follows,

$$\min_{x_0,\ldots,x_{N_T-1}} R(x_0,\ldots,x_{N_T-1}), \quad \text{subject to:} \quad D(x_0,\ldots,x_{N_T-1}) \leq D_{max}, \tag{5.15}$$

where D_{max} is the maximum energy permissible in the DFD. As we pointed out in the section 5.1, we could also use the optimal bit allocation scheme we proposed in section 4.3 for dependent quantizers for the minimum maximum distortion objective. Since, however, it is more common to minimize the total DFD energy than to minimize the maximum DFD energy, the proposed scheme is based on the minimum total distortion bit allocation scheme.

5.2.3 Recursive generation of admissible motion vector sets

From a theoretical point of view, every possible motion vector of block $b_{l,i}$ should be included in the set $M_{l,i}$, which is the admissible motion vector set for block $b_{l,i}$. This means that for a typical search window of ± 15.5 pixels and an accuracy of 0.5 pixel, $|M_{l,i}|$ = 63*63 =3969, which is quite large. Most of these motion vectors, however, are not likely candidates for the optimal path, since they do not correspond well to the real motion in the scene and therefore they lead to a high distortion and a high rate. These motion vectors can be found by performing block matching since they will result in a high matching error.

To make the optimization process faster, the prior knowledge about these motion vectors, which are unlikely candidates for the optimal motion vector, should be taken into account. Even though this might be complicated in general, it can be easily achieved in the presented framework of dynamic programming by reducing the set $M_{l,i}$ of admissible motion vectors of block $b_{l,i}$.

We propose the following scheme for the recursive generation of admissible motion vector sets. An initial motion vector search is conducted for the $2^{n_0} \times 2^{n_0}$ blocks at level n_0 by doing block matching with integer accuracy. The K integer motion vectors which lead to the best prediction are kept. Then the set $M_{n_0,i}$

is defined as the set which contains the K top integer motion vectors plus their half pixel neighbors.

After the set of admissible motion vectors has been defined for the bottom level (n_0), the sets of admissible motion vectors for higher level blocks can be defined recursively. A block $b_{l,i}$ only includes a motion vector in $M_{l,i}$ if the motion vector has been selected by all of its child nodes. This leads to the fact that for small blocks, many motion vectors are considered but the bigger the blocks get, the fewer motion vectors are included. This reflects the well known fact that small block sizes lead to small energy in the DFD but not very consistent motion vector fields, whereas bigger blocks lead to consistent vector fields, but the energy in the DFD can be quite high [16]. Our experiments have shown that for $K = 10$ this restriction of the search space does not lead to a performance loss but increases the speed of the algorithm significantly.

5.2.4 Example

In this example we compare a QT-based optimal motion estimation scheme with the fixed block size based scheme presented in section 5.1. The QT-based optimal motion estimator uses the same scheme to encode the DVF and the same 16×16 modified Hilbert scan as the optimal fixed block size based motion estimator. Since the smallest block size in the proposed QT-based optimal motion estimator is 8×8, the modified Hilbert scan is developed recursively, as proposed in section 4.4.3, down to level $n_0 = 3$ as displayed in Fig. 5.5. Again, for the presented experiment, we use the 176-th and 180-th frames of the QCIF sequence "Mother and Daughter" which are displayed in Fig. 5.1. We only use the luminance part of the sequence to perform the presented experiments, but an extension which includes the chrominance can easily be accomplished. We intra code the 176-frame using TMN4 and a quantizer step size of 10. The resulting PSNR for this frame is 33.85 dB . Then the original 180-th frame is used to find the DVF. First we employ the TMN4 block matching scheme which uses the sum of the absolute error and favors the zero motion vector by reducing its error by a constant amount of 100. The resulting bit rate for the DVF and the DFD energy are listed in row "TMN4 (16×16)" in Table 5.1. The predicted frame and the DVF are displayed in Fig. 5.2.

Now the QT-based optimal motion estimator is run, where first the maximum rate is set equal to the TMN4 rate, i.e., $R_{max} = 470$ bits. We call this experiment "matched Rate (QT)" in Table 5.1 and the resulting rate and distortion are listed in the corresponding row. Note that for the same rate, the total distor-

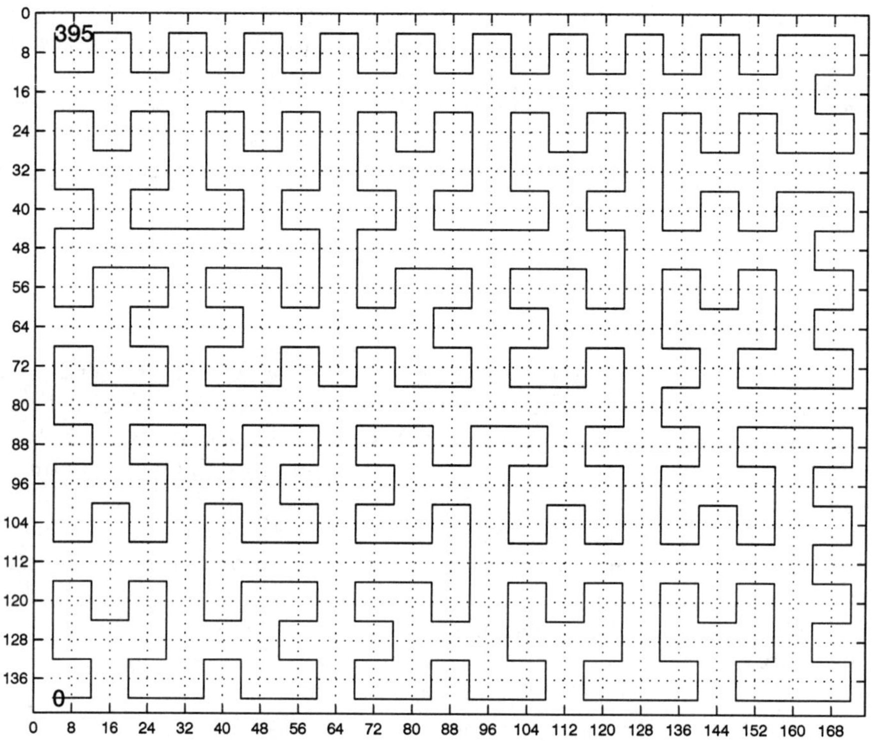

Figure 5.5 Modified Hilbert scan for level $n_0 = 3$ of a QCIF frame.

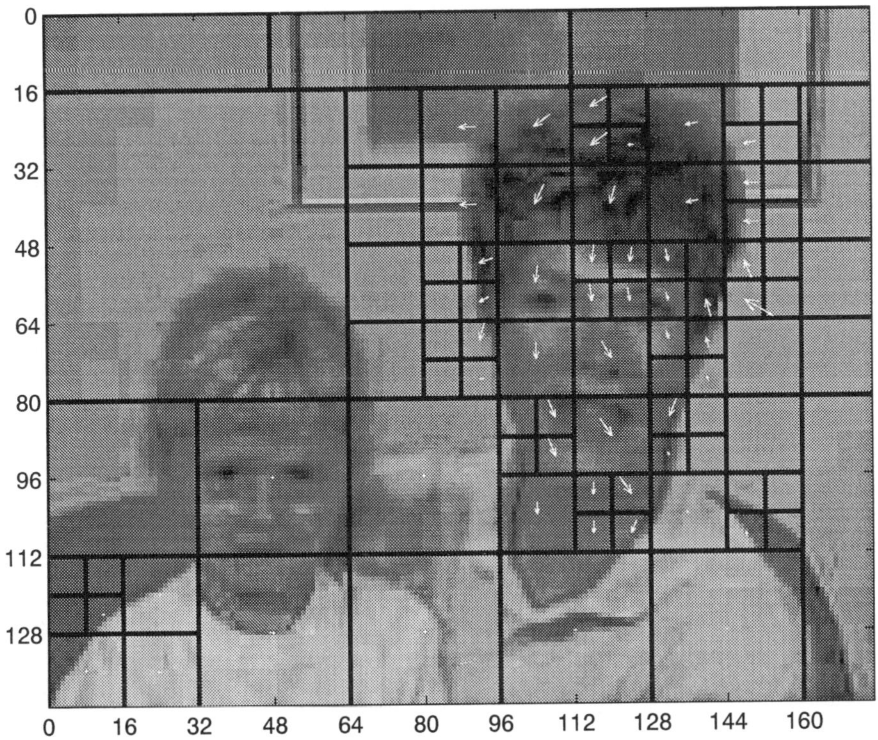

Figure 5.6 The predicted frame and the DVF for the optimal QT-based block matching algorithm, when the rate matches the TMN4 rate.

tion is reduced significantly, or in other words, a better prediction is achieved. Besides outperforming the TMN4 motion estimator, the proposed QT-based scheme is also better than the optimal block matching scheme for a fixed block size of 16×16. This is due to the more compact representation of the motion by means of the QT, which enables the QT-based motion estimator to spend more bits in ares with complex motion, i.e., small block sizes are used, and less bits in areas with simple motion, i.e., large block sizes are used. The better prediction performance is also apparent from the resulting predicted frame and the DVF which are displayed in Fig. 5.6.

Recall that the n_0 level is scanned by the modified Hilbert scan shown in Fig. 5.5. To generate the scanning path for an arbitrary QT decomposition, the

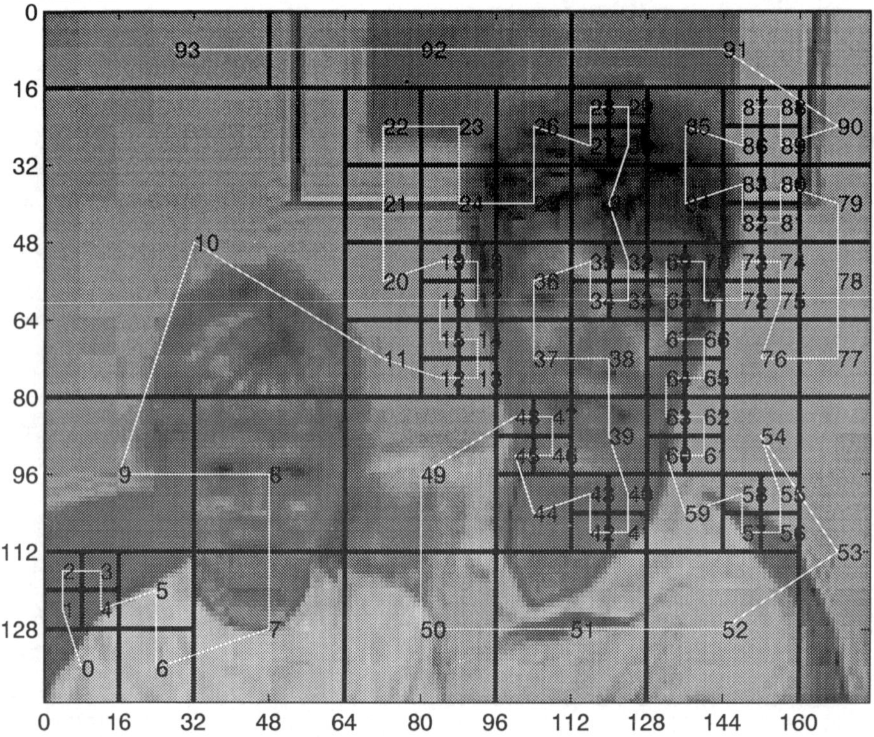

Figure 5.7 The overall scanning path.

procedure introduced in section 4.11 is used. In Fig. 5.7 the resulting overall scanning path which corresponds to the QT decomposition displayed in Fig. 5.6 is shown.

For the next experiment, the DFD energy of the TMN4 run is matched by setting D_{max} =30.65 dB PSNR. We call this experiment "matched Dist. (QT)" in Table 5.1 and the resulting rate and distortion are listed in the corresponding row. Note that for the same distortion, the bit rate is reduced significantly, and again the bit rate is also lower than that of the optimal fixed block size based scheme. The resulting predicted frame and the DVF are displayed in Fig. 5.8. Again, even though the DFD energy is the same as in Fig. 5.2, the predicted frame is of higher quality. The same explanation applies as before, that is,

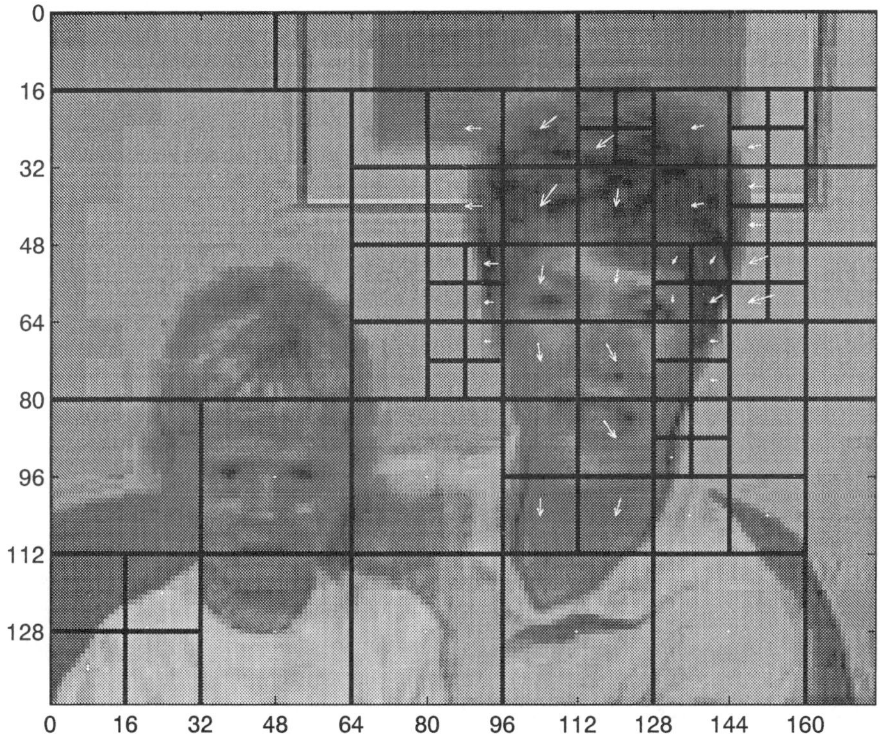

Figure 5.8 The predicted frame and the DVF for the optimal QT-based block matching algorithm, when the distortion matches the TMN4 distortion.

the implicit DVF smoothing along the Hilbert scan results in a good DVF. In addition, the inhomogeneous structure of the DVF fits a real DVF better and hence a better representation of the real DVF can be achieved. We will use the fact that this rate constrained motion estimator results in smooth motion fields in section 5.3. In it we propose a scheme which uses this motion estimator for motion compensated interpolation, where the scheme offers complete control of the optimal tradeoff between interpolation error energy and DVF smoothness.

5.3 MOTION COMPENSATED INTERPOLATION

In this section we propose a scheme for motion compensated interpolation which is based on the optimal motion estimator presented in the previous section [181]. Besides being a very general problem in digital video research, motion compensated interpolation can be used in the context of very low bit rate video coding to increase the displayed frame rate. In very low bit rate coding, it is very common to down-sample the original sequence in the temporal direction. In fact the most popular frame rates are 7.5 and 10 frames per second, or in other words, only every fourth or third frame of the original 30 frames per second sequence is encoded.

The down-sampling results in some jerkiness since the human observer starts to notice the single frames at such low frame rates. The displayed frame rate can be increased by interpolating missing frames using motion compensated interpolation. The problem with motion compensated interpolation is that if an error is made, it will be clearly visible in the sequence since human observers are very sensitive to irregularities along the temporal direction. As pointed out before, the above motion estimator results in a motion vector field which is very close to the expected field. It therefore can be used for motion compensated interpolation.

5.3.1 Regular grid approach

The idea behind the proposed motion compensated interpolation scheme can be explained by considering Fig. 5.9. In it the left most frame $\hat{f}_0$ and the right most frame $\hat{f}_N$ are the transmitted and reconstructed frames at the decoder. The frames between these two frames are the ones we wish to interpolate. In Fig. 5.9, frame 1 is the frame which is being interpolated. The most common

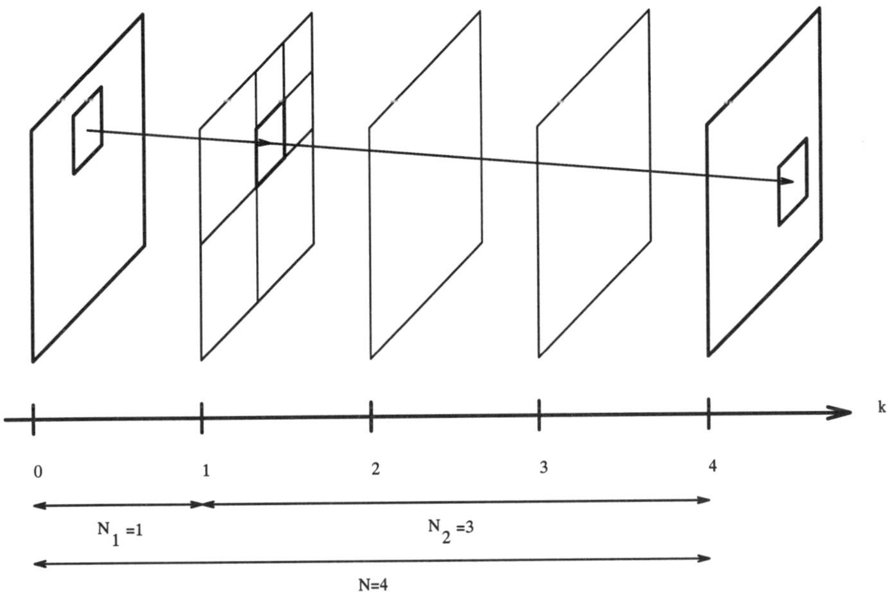

Figure 5.9 Motion compensated interpolation of the first frame using the two transmitted frames, which are the left most and the right most frame.

approach for motion compensated interpolation is to find the DVF between the left most and right most frames and then to use this motion information to interpolate the frames in between. One problem with this approach is that the DVF needs to be interpolated for the interpolated frames, since one cannot guarantee that the DVF will fall onto grid points in these frames. In the proposed approach, we avoid the problem of an irregular grid since the frame which has to be interpolated is segmented (instead of the left most or right most frame as in traditional approaches) using a QT decomposition and the DVF is found for this frame and hence it will by default be on grid points.

5.3.2 Motion compensated interpolation

As is evident from Fig. 5.9, we assume non-accelerated motion (also called linear motion since the trajectory is a straight line). By segmenting the frame which has to be interpolated, we can assign a motion vector m_j to each block $b_{l,i}$ in the segmented frame. The motion vector is with respect to the previous frame in the sequence, even if the previous frame is not one of the reconstructed frames. This will become clearer later on. The QT segmentation and the motion vector, which are found using a procedure described in section 5.3.3, are then used to interpolate block $b_{l,i}$ of the k-th frame as follows,

$$b_{l,i}(x,y) = \frac{N_2 \hat{f}_0(x - N_1 m_{j,x}, y - N_1 m_{j,y}) + N_1 \hat{f}_N(x + N_2 m_{j,x}, y + N_2 m_{j,y})}{N_1 + N_2} \quad \forall\, (x,y) \in b_{l,i}, \tag{5.16}$$

where $m_{j,x}$ and $m_{j,y}$ are the x and y coordinate of the motion vector m_j of block $b_{l,i}$. N_1 and N_2 are the temporal distances of the frame which is being interpolated from frames $\hat{f}_0$ and $\hat{f}_N$, respectively (see Fig. 5.9). This weighted interpolation leads to a fading effect so that the transmitted frame closer to the interpolated frame has a higher weight than the transmitted frame which is farther away.

5.3.3 Operational rate distortion formulation

Using the above interpolation formula, it is clear that the quality of the interpolated frame only depends on the QT segmentation and the DVF. As we pointed out in the previous section, the QT-based motion estimator results in very good DVFs and hence we would like to employ it to find the QT segmentation and the DVF such that the motion compensated interpolation results in a visually pleasing sequence. To that end we define a block interpolation error (distortion) of block $b_{l,i}$ using motion vector m_j (which is contained in node x_j) as follows,

$$\begin{aligned} d_j(x_j) \;=\; & \sum_{(x,y)\in b_{l,i}} \Big(\hat{f}_0(x - N_1 * m_{j,x}, y - N_1 * m_{j,y}) \\ & - \hat{f}_N(x + N_2 * m_{j,x}, y + N_2 * m_{j,y})\Big)^2 . \end{aligned} \tag{5.17}$$

This formula expresses the fact that the scheme looks for the same block in $\hat{f}_0$ and $\hat{f}_N$ which must pass through the block under consideration in the interpolated frame. This is also indicated in Fig. 5.9. As we pointed out before, the motion vector m_j is with respect to the next frame in the sequence, hence it needs to be scaled by the temporal distances N_1 and N_2 in the above formula. Again the frame distortion is defined as the sum of the distortions of the blocks used in the QT decomposition, hence,

$$D(x_0, \ldots, x_{N_T-1}) = \sum_{j=0}^{N_T-1} d(x_j). \tag{5.18}$$

Intuitively it seems clear that if we find a QT and a DVF such that the frame distortions is minimized, then the interpolated frame should be as good as possible under the linear motion assumption. This, however, is not the case since the smallest frame distortion can be achieved by using one motion vector for each of the smallest possible blocks in the QT decomposition. These motion vectors would be selected so that the individual block distortions are minimized, which in turn results in the smallest frame distortion. The resulting DVF would be quite noisy, since the ultimate goal is to minimize the above distortion. Clearly, minimizing the frame distortion will not result in a good interpolation, since we did not take into account the smoothness of the resulting DVF. It is well known that a real DVF is quite smooth, except at object boundaries and in uncovered regions. Therefore it is common in motion estimation not to minimize the DFD energy alone, but to regularize the problem by adding a term to the objective function which measures the smoothness of the DVF. This can be easily achieved in the proposed framework, since the rate necessary to transmit the DVF is known by the QT-based motion estimator. As we have pointed out before, a smooth DVF results is a small rate, whereas a noisy DVF requires a high bit rate. This follows from the first oder DPCM employed to encode the DVF. Additionally, the Hilbert scan enforces the smoothness of the DVF in all dimensions, which is one of the advantages of a Hilbert scan over a raster scan. Therefore we propose to find the DVF and the QT decomposition as the solution of the following regularized optimization problem,

$$\min_{x_0,\ldots,x_{\mathcal{T}-1}} (D(x_0,\ldots,x_{\mathcal{T}-1}) + \lambda * R(x_0,\ldots,x_{\mathcal{T}-1})), \tag{5.19}$$

where λ is the regularization parameter. Clearly, this is exactly the same optimization problem which we have been solving iteratively for different λs to find the optimal DVF and QT decomposition for the optimal QT-based motion estimator. Hence we can find the optimal solution to this regularized optimization problem, which is very uncommon in motion estimation, where a local optimum is usually all that can be found.

5.3.4 Example

We apply the proposed motion compensated interpolation scheme to two compressed QCIF black and white sequences, "Miss America" and "Mother and Daughter". These sequences have been compressed by a video compression scheme which we will present in chapter 7. In this scheme it is possible to fix the frame distortion for each frame, and we set it to 34.0 dB PSNR for the "Mother and Daughter" sequence and 35.0 dB PSNR for the "Miss America" sequence. The compressed frame rate of these sequences is 15 frames per second (fps) or in other words, only every second frame of the original sequence is encoded. In very low bit rate coding, it is common to skip as many as four frames, which results in a frame rate of 7.5 fps. At such low frame rates, a human observer starts to notice the single frames, which can be very disturbing. For the presented experiments, we down-sample these two 15 fps sequences temporally by skipping every second frame. Clearly, this results in two new sequences with frame rates of 7.5 fps. We use the proposed motion compensated interpolation scheme to increase this frame rate back to 15 fps, or in other words, we interpolate the frames we skipped during the down-sampling. We then compare the reconstructed frames with the interpolated frames visually and using the PSNR between the frames and the original frame.

For the implementation of the proposed motion compensated interpolation algorithm, the same QT-based motion estimator is used as in the experiments presented in the previous section. Based on experiments, we select a λ of 0.01. For a discussion on how to select an appropriate λ in the context of image restoration, see [56]. In addition, the smallest level of the QT are 4×4 blocks, hence n_0 is equal to 2, which is different from the QT based optimal motion estimator presented in the previous section where n_0 was equal to 3.

Both sequences, when interpolated to 15 fps, are more pleasing to a human observer than the 7.5 fps sequences, since they are less jerky. To illustrate the interpolation results, we display a sequence of three reconstructed frames, i.e., consecutive frames of the 15 fps sequences and compare them with the same frames resulting from the interpolation scheme. Clearly the first and last frame are identical, whereas the interpolated middle frame should be as close as possible to the middle frame of the reconstructed sequence.

In Fig. 5.10 the 84-th, 86-th and 88-th reconstructed frames of the "Miss America" sequence are displayed. In Fig. 5.11 the 84-th and 88-th reconstructed frames of the "Miss America" sequence are displayed on the left and on the right, respectively. The 86-th frame is interpolated using the proposed motion compensated interpolation scheme with $\lambda = 0.01$. Clearly the interpolated frame is very similar to the encoded frame. Recall that the interpolation is done at the decoder and is completely independent from the encoder. In other words, no additional information needs to be sent. The PSNR of the interpolated frame is 34.0 dB which is one dB lower than that of the reconstructed frame. On the other hand, to transmit the reconstructed frame, 1264 bits where necessary. In Fig. 5.12 the QT segmentation and the DVF are displayed together with the interpolated frame.

Figure 5.10 The 84-th, 86-th and 88-th reconstructed frames of the "Miss America" sequence. The PSNR of each frame is 35.0 dB.

Figure 5.11 The 84-th and 88-th reconstructed frames of the "Miss America" sequence are displayed on the left and on the right, respectively. The frame in the middle is the interpolated 86-th frame, its PSNR is 34.0 dB.

In Fig. 5.13 the 176-th, 178-th and 180-th reconstructed frames of the "Mother and Daughter" sequence are displayed. In Fig. 5.14 the 176-th and 180-th reconstructed frames of the "Mother and Daughter" sequence are displayed on the left and on the right, respectively. The 178-th frame is interpolated using the proposed motion compensated interpolation scheme with $\lambda = 0.01$. Again, the interpolated frame is very similar to the encoded frame. The PSNR of the interpolated frame is 33.3 dB which is 0.7 dB lower than that of the reconstructed frame. On the other hand, to transmit the reconstructed frame, 3254 bits where necessary. In Fig. 5.15 the QT segmentation and the DVF are displayed together with the interpolated frame.

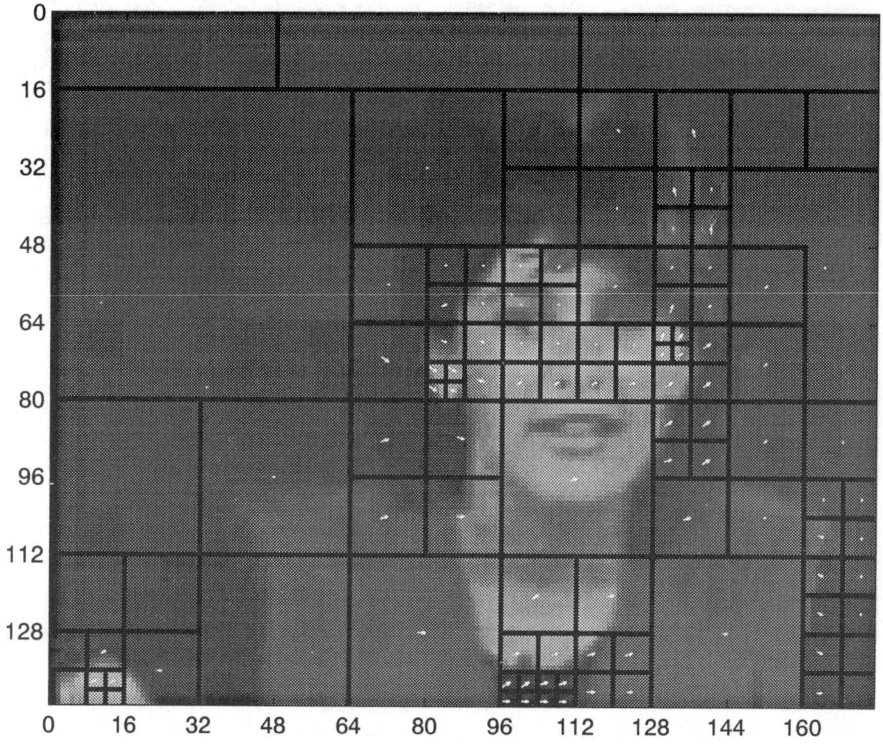

Figure 5.12 The QT segmentation and the DVF of the 86-th interpolated frame of the "Miss America" sequence.

5.4 SUMMARY

In this chapter we presented two motion estimators and a motion compensated interpolation scheme. Both motion estimators are optimal in an operational rate distortion sense. The first motion estimator is region-based and the DVF can be encoded using any order of dependencies, but clearly the lower the order is, the more efficient the optimization approach. The second motion estimator is QT-based and the DVF is encoded using a first order DPCM scheme. Both motion estimators outperform the commonly used block matching algorithms, and the QT-based motion estimator has the highest performance since the inhomogeneous representation of the DVF is very efficient. The QT-based motion estimator is then used in a motion compensated interpolation scheme,

Figure 5.13 The 176-th, 178-th and 180-th reconstructed frames of the "Mother and Daughter" sequence. The PSNR of each frame is 34.0 dB.

Figure 5.14 The 176-th and 180-th reconstructed frames of the "Mother and Daughter" sequence are displayed on the left and on the right, respectively. The frame in the middle is the interpolated 178-th frame, its PSNR is 33.3 dB.

where the interpolated frame is found using the DVF and the QT decomposition found by the QT motion estimator. The QT motion estimator minimizes the interpolation error energy to which a regularization term is added, so that a smooth DVF results. We showed that this regularization problem is identical to the minimization of the relaxed problem for the Lagrangian multiplier method, and hence the solution can be found efficiently by dynamic programming.

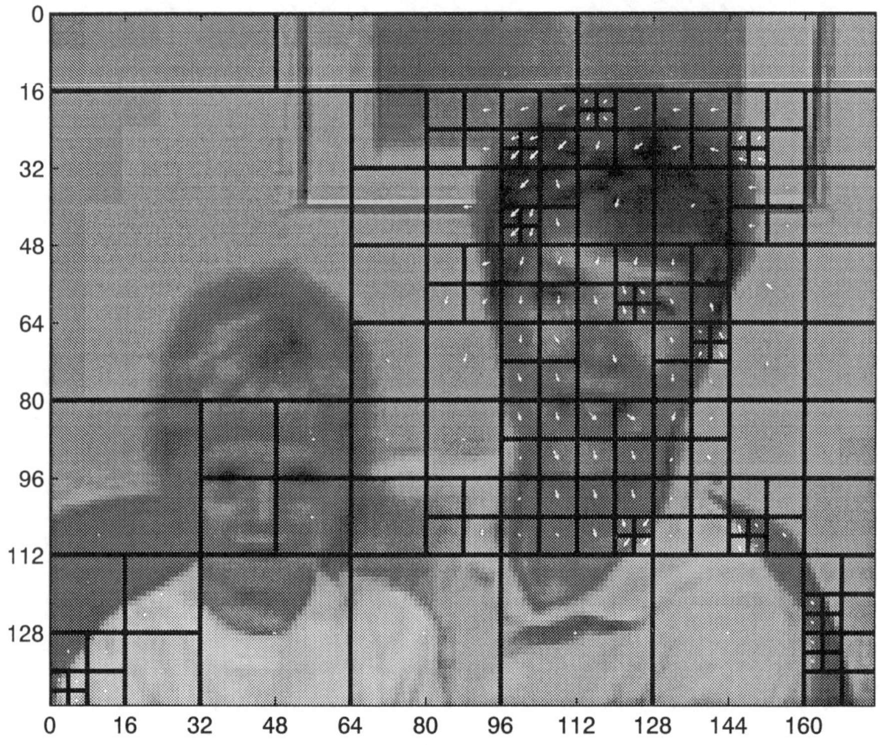

Figure 5.15 The QT segmentation and the DVF of the 178-th interpolated frame of the "Mother and Daughter" sequence.

6

A VIDEO COMPRESSION SCHEME WITH OPTIMAL BIT ALLOCATION BETWEEN DISPLACEMENT VECTOR FIELD AND DISPLACED FRAME DIFFERENCE

In this chapter we introduce a theory for the optimal bit allocation between displacement vector field (DVF) and displaced frame difference (DFD) for region-based motion compensated video coders [184, 185]. The proposed scheme is based on the optimal region matching algorithm presented in section 5.1. The optimal region matching algorithm finds the DVF which results in the smallest DFD energy for a given maximum rate for encoding the DVF. In this chapter we extend this scheme to include the quantizer selection so that the resulting distortion of the reconstructed frame is minimized for a given total bit rate. As before, the DVF is encoded which results in the DVF bit rate, but now the quantized DFD is also encoded and hence the total bit rate is the sum of the DVF rate and the DFD rate.

We first consider the case of a lossless motion compensated video coder (MCVC) and derive a general Dynamic Programming (DP) formulation which results in an optimal tradeoff between the DVF and the DFD. We then consider the more important case of a lossy MCVC and present an algorithm which solves the tradeoff between the rate and the distortion. This algorithm is based on Lagrangian relaxation and the DP approach introduced for the lossless MCVC. We then present an H.263-based MCVC which uses the proposed optimal bit allocation and compare its results to H.263. As expected, the proposed coder is superior in the rate distortion sense. In addition to this, it offers many advantages for a rate control scheme. The presented theory can be applied to build new optimal coders and to analyze the heuristics employed in existing coders. In fact whenever one changes an existing coder, the proposed theory can be used to evaluate how the change affects its performance.

6.1 INTRODUCTION

A fundamental problem in every motion compensated video coder (MCVC) is the bit allocation between the DFD and the DVF. In this chapter we present a general framework which uses operational rate distortion theory to solve this problem for a finite set of admissible quantizers and motion vectors.

There have been previous attempts to find an optimal tradeoff between DVF and DFD. In the standards, such as MPEG-1, MPEG-2, H.261 and H.263, the bit allocation between DFD and DVF is not explicitly defined and there is great freedom for selecting the motion vectors and the quantizers. Block matching is the most common approach for finding the DVF. The resulting DFD is encoded with quantizers selected by a rate control algorithm. In such a scheme, the tradeoff between DVF and DFD is implicit, without taking into account the resulting rate and distortion.

In [123], the authors assume a stochastic model for the distribution of the DFD and proceed to calculate the entropy of a given block based on some observed statistics. This entropy is then used to decide if a block should be split into four smaller blocks with their own motion vectors, or if the block should be kept as a basic unit.

In [67], the problem of rate constrained motion estimation is considered and the optimal bit allocation condition for a strictly convex and everywhere differentiable multivariate rate distortion function is derived. It is applied to the problem of optimal bit allocation between the DVF and the DFD and a rate constrained, region-based motion estimator is introduced. In this chapter, we do not assume knowledge of a convex and everywhere differentiable multivariate rate distortion function, but instead we deal with a set of finite quantizers and motion vectors.

In [96], a variable block size motion estimator is presented. It is implied that the motion vectors are encoded by pulse code modulation (PCM) and hence the resulting optimization procedure is quite simple and in fact equivalent to the one presented in [200]. In contrast to this work, we allow for more sophisticated encoding schemes for the DVF, such as the popular differential PCM (DPCM). This leads to a more complex optimization problem for which we derive the optimal solution.

In [168], the optimal bit allocation problem for lossless video coders is studied. The authors use a stochastic model which has been derived in [123] to find a formula for the entropy of the DFD as a function of the DVF accuracy. A similar formula can also been found in [65, 66]. The stochastic model derived in [123] cannot be applied if the DFD is encoded by a sophisticated encoding scheme.

The main contribution of this chapter is that the theory we present is very general and allows for the optimization of a wide range of schemes. Furthermore, the presented theory allows the rate and the distortion of a given region to depend on quantizers and motion vectors of other regions. This enables efficient encoding of the DVF using a predictive scheme, such as DPCM, and the use of distortion measures which also include region boundary effects which are very important for human observers.

Since we work with operational rate distortion functions, we do not assume any stochastic models or convexity properties. Stochastic models are commonly used to estimate the entropy of the prediction error, but our experiments have shown that when a sophisticated encoding scheme is used, such as discrete cosine transform (DCT) and run-length encoding, these estimates can be quite inaccurate. The convexity and differentiability properties usually invoked for the rate distortion function imply a continuous function. However, in every real video coder the set of motion vectors and quantizers is finite and hence the operational rate distortion curves are neither differentiable everywhere nor necessarily convex. It is commonly assumed or implied that the motion vectors are encoded by a simple PCM scheme which is inefficient, since neighboring motion vectors are highly correlated and a DPCM scheme is more appropriate. The problem with using DPCM for the encoding of the motion vectors is the dependency it introduces in the optimization procedure which cannot be solved by any of the previously proposed approaches. As we will show, these dependencies can be handled efficiently in the proposed framework. The commonly used distortion measures are intra region based. In other words, these measures do not capture the boundary effects, such as blocking artifacts. It is well known that these artifacts are highly visible to a human observer. Again, the proposed framework does allow for such measures.

This chapter is organized as follows: In section 6.2, we define the problem under consideration. In section 6.3 we derive the optimal solution for lossless MCVCs. This solution is then extended in section 6.4 to include lossy MCVCs which are optimal for the minimum total distortion criterion and in section 6.5 the lossless solution is extended to include lossy MCVCs which are optimal for the minimum maximum distortion criterion. In section 6.6 we develop a lossy video coder based on the presented theory and in section 6.7 we discuss some implementation issues which reduce the computational complexity of the presented coder. The experimental results of this coder are presented in section 6.8 and this chapter is summarized in section 6.9.

6.2 NOTATION AND ASSUMPTIONS

In this section we introduce the necessary notation and state the assumptions which will be used in the rest of the chapter. Note that the notation in this chapter is very similar to the one introduced in section 5.1.1. Our study of the optimal bit allocation between the DVF and DFD is restricted to the frame level. In other words, we do not attempt to optimally allocate the bits among the different frames of a video sequence. The reader interested in that problem is referred to [163]. For the rest of this chapter we assume that a rate control algorithm has given us the maximum number of bits available (R_{max}) or the maximum acceptable distortion (D_{max}) for each frame.

We assume that the current frame is segmented into N regions, $o_0, \ldots, o_{N-1}$, and that this segmentation and the associated scanning path are known to both the encoder and the decoder. We then number the regions such that the scanning path visits them in ascending order. Every region o_i has a motion vector $m_i \in M_i$, and a quantizer $q_i \in Q_i$ associated with it, where M_i is the set of all admissible motion vectors for region o_i and Q_i is the set of all admissible quantizers for region o_i. As in every practical video coding scheme, we assume that the sets M_i and Q_i are finite. Let us define a decision vector $v_i = [m_i, q_i] \in V_i$, for every region o_i which contains the motion vector and the quantizer for that region. $V_i = M_i \times Q_i$ is the admissible decision vector set for region o_i.

In a motion compensated video coder, $D_k(v_0, \ldots, v_{N-1})$, the distortion of frame k, is a function of the previously reconstructed frame $\hat{f}_{k-1}$, the DVF $= \{m_i\}_{i=0}^{N-1}$, and the current frame f_k. In addition, we assume that the DVF rate $R_k(v_0, \ldots, v_{N-1})$ is a function of $\hat{f}_{k-1}$, f_k and the DVF.

The next assumption expresses the idea that the frame rate and frame distortion can be decomposed into a sum of region rates $r_i(v_{i-a}, \ldots, v_{i+b})$ and region distortions $d_i(v_{i-a}, \ldots, v_{i+b})$, which only depend on a local neighborhood. We assume that there exist integers $a \geq 0$ and $b \geq 0$, and a family of functions d_i

and r_i such that the following holds,

$$D_k(v_0, \ldots, v_{N-1}) = \sum_{i=0}^{N-1} d_i(v_{i-a}, \ldots, v_{i+b}), \tag{6.1}$$

$$R_k(v_0, \ldots, v_{N-1}) = \sum_{i=0}^{N-1} r_i(v_{i-a}, \ldots, v_{i+b}), \tag{6.2}$$

where the decision vectors v_j not belonging to any region ($j \notin [0, \ldots, N-1]$), represent the boundary parameters and can be set to any desired value. The above assumption is very important since the efficiency of the optimization procedure to be introduced later will directly depend on $a+b+1$ which defines the size of the neighborhood.

It is noted here that assumptions (6.1) and (6.2) are quite general and valid for every existing video coding standard. One contribution of this chapter is the formulation of the optimal bit allocation problem for a MCVC. It is important to realize that assumptions (6.1) and (6.2) are essential for the development which follows. Most commonly used distortion measures, such as the mean squared error (MSE) or the peak signal to noise ratio (PSNR), satisfy assumption (6.2). For example, in MPEG-1 and MPEG-2, the rate for a given block depends not only on the quantizer and motion vector of that block, but also on the motion vector of the previous block, which is used as a predictor for the current motion vector. Usually the block distortion measured by the popular MSE measure depends only on the motion vector and the quantizer of the current block. A noteworthy exception is H.263 when the "Advanced Prediction Mode" is used. Then, overlapped block motion compensation is employed and the MSE of a given 8×8 block depends on four spatial neighbors.

6.3 LOSSLESS MCVC

In section 5.1.2 we showed that the minimization of the DFD energy, subject to a maximum DVF rate, can be accomplished using the general concept of dependent quantization introduced in section 4.1. It is possible to convert the optimal bit allocation problem between DVF and DFD into the same general form used in section 4.1, and hence the algorithm proposed in section 4.1 can be applied to find the optimal solution.

In this chapter we take a slightly different route, in that we first derive the optimal bit allocation for a lossless scheme explicitly and then show how this solution can be used, in the Lagrangian relaxation step, to solve the optimal bit allocation problem for a lossy MCVC. Note that this is the same approach we used in section 4.5 to derive the optimal quad-tree decomposition for quad-trees with leaf dependencies. This approach has the advantage that the optimal solution of the lossless case is also found and it clearly shows the separation between the optimization algorithm (DP) and the Lagrangian relaxation step.

As mentioned above, in this section we study the case of a lossless MCVC. Since the reconstructed frame is identical to the original frame, the frame distortion will be zero and the goal is to minimize the number of bits required for the DVF and the DFD. This can be stated as follows,

$$\min_{v_0,\ldots,v_{N-1}} R_k(v_0,\ldots,v_{N-1}). \tag{6.3}$$

Since this is a lossless MCVC, the DFD is not quantized, but encoded losslessly. With a slight abuse of notation let q_i represent different lossless encoding schemes for region o_i (i.e., DPCM with different predictor order, etc.), instead of different quantizers. Since we will refer to this algorithm later, we will still call the q_i's quantizers in the following derivation.

Since we deal with a finite number of admissible motion vectors and quantizers, the above optimization problem can clearly be solved by an exhaustive search. The time complexity for such an exhaustive search is $O(|V_i|^N)$, where $|V_i|$ denotes the cardinality of V_i, and we assume that all V_is have the same cardinality. We will show that the proposed algorithm reduces this complexity significantly. Note that when we use the term time complexity, we refer to the number of comparisons necessary to find the optimal solution. This does not include the time spent to evaluate the operational rate distortion functions, since this strongly depends on the implementation of a given MCVC.

As stated in assumption (6.2), the frame rate R_k is the sum of rates which only depend on local neighborhoods. We will now employ this assumption to derive a dynamic programming solution to problem (6.3).

We use generic terms, ($g_i(v_{i-a},\ldots,v_{i+b})$ and $G_k(v_0,\ldots,v_{N-1})$), in this derivation since we will refer to them later. Let

$$g_i(v_{i-a},\ldots,v_{i+b}) = r_i(v_{i-a},\ldots,v_{i+b}), \tag{6.4}$$

$$G_k(v_0,\ldots,v_{N-1}) = \sum_{i=0}^{N-1} g_i(v_{i-a},\ldots,v_{i+b}), \tag{6.5}$$

and

$$g_l^*(v_{l-a+1},\ldots,v_{l+b}) = \min_{v_0,\ldots,v_{l-a}} \sum_{i=0}^{l} g_i(v_{i-a},\ldots,v_{i+b}). \tag{6.6}$$

From Eq. (6.6) it follows that,

$$\begin{aligned} g_{l+1}^*&(v_{l+1-a+1},\ldots,v_{l+1+b}) \\ &= \min_{v_0,\ldots,v_{l+1-a}} \sum_{i=0}^{l+1} g_i(v_{i-a},\ldots,v_{i+b}) \qquad (6.7) \\ &= \min_{v_{l+1-a}} \left[\min_{v_0,\ldots,v_{l-a}} \left(\sum_{i=0}^{l} g_i(v_{i-a},\ldots,v_{i+b}) + g_{l+1}(v_{l+1-a},\ldots,v_{l+1+b}) \right) \right] \qquad (6.8) \end{aligned}$$

Since $g_{l+1}(v_{l+1-a},\ldots,v_{l+1+b})$ does not depend on $v_0,\ldots,v_{l-a}$, it can be moved outside the inner minimization. Then the resulting inner minimization is equal to $g_l^*(v_{l-a+1},\ldots,v_{l+b})$ in Eq. (6.6) and the following DP recursion formula results,

$$\begin{aligned} g_{l+1}^*&(v_{l+1-a+1},\ldots,v_{l+1+b}) \\ &= \min_{v_{l+1-a}\in V_{l+1-a}} \left[g_l^*(v_{l+1-a},\ldots,v_{l+b}) + g_{l+1}(v_{l+1-a},\ldots,v_{l+1+b}) \right]. \qquad (6.9) \end{aligned}$$

Forward DP (also called the Viterbi algorithm [53]) can now be used to solve problem (6.3). First we need to initialize the recursion which is achieved in the following way,

$$\begin{aligned} g_{a-1}^*(v_0,\ldots,v_{a+b-1}) &= \sum_{i=0}^{a-1} g_i(v_{i-a},\ldots,v_{i+b}) \\ &\qquad \forall\, [v_0,\ldots,v_{a+b-1}] \in V_0 \times \ldots \times V_{a+b-1}. \qquad (6.10) \end{aligned}$$

Next, the recursion is started, and the DP recursion formula (6.8) is applied for $l = a-1$ up to and including $l = N-2$. That is,

$$\begin{aligned} g_{l+1}^*&(v_{l+1-a+1},\ldots,v_{l+1+b}) \\ &= \min_{v_{l+1-a}\in V_{l+1-a}} \left[g_l^*(v_{l+1-a},\ldots,v_{l+b}) + g_{l+1}(v_{l+1-a},\ldots,v_{l+1+b}) \right], \\ &\quad \forall\, [v_{l+1-a+1},\ldots,v_{l+1+b}] \in V_{l+1-a+1} \times \ldots \times V_{l+1+b}. \qquad (6.11) \end{aligned}$$

Then the final solution is found by observing that,

$$\min_{v_0,\ldots,v_{N-1}} G(v_0,\ldots,v_{N-1}) = \min_{v_{N-a},\ldots,v_{N-1}} g_{N-1}^*(v_{N-a},\ldots,v_{N-1+b}). \tag{6.12}$$

As we have seen the time complexity for the exhaustive search is exponential. The time complexity for the DP approach depends directly on the size of the neighborhood and is $O(N * |V_i|^{a+b+1})$, where again we assume that all V_is have the same cardinality.

6.3.1 Example

We now consider a simple example to illustrate the above theory. Assume that a lossless MCVC has an admissible motion vector set $M_i = \{ma, mb\}$, and two different lossless encoding schemes, i.e., $Q_i = \{qa, qb\}$. Let the entire frame be split into $N = 4$ regions and the motion vectors be encoded using a first order DPCM along the scanning path. Using these definitions, the goal is to minimize the frame rate $R_k(v_0, v_1, v_2, v_3)$ with respect to the motion vector and quantizer choices. First we have to identify the size of the neighborhood involved in this problem. Since a first order DPCM along the scanning path is used for the encoding of the motion vectors, only the previous motion vector is required for determining the rate of the current region. Therefore $a = 1$ and $b = 0$.

We can now use Eq. (6.9) to initialize the forward DP, i.e.,

$$g_0^*(v_0) = g_0(v_{-1}, v_0), \qquad \forall\, v_0 \in V_0, \tag{6.13}$$

where v_{-1} can be set to any value, for example $v_{-1} = [ma, qa]$. Next, the recursion is started, that is, the DP recursion formula (6.8) is applied. First for $l = 0$,

$$g_1^*(v_1) = \min_{v_0} \left[g_0^*(v_0) + g_1(v_0, v_1)\right], \qquad \forall\, v_1 \in V_1, \tag{6.14}$$

then for $l = 1$,

$$g_2^*(v_2) = \min_{v_1} \left[g_1^*(v_1) + g_2(v_1, v_2)\right], \qquad \forall\, v_2 \in V_2, \tag{6.15}$$

and finally,

$$g_3^*(v_3) = \min_{v_2} \left[g_2^*(v_2) + g_3(v_2, v_3)\right], \qquad \forall\, v_3 \in V_3. \tag{6.16}$$

The final solution can now be found using Eq. (6.12), that is,

$$\min_{v_0, v_1, v_2, v_3} G_k(v_0, v_1, v_2, v_3) = \min_{v_3} g_3^*(v_3). \tag{6.17}$$

A good tool to help visualize DP is a trellis. In Fig. 6.1 the trellis corresponding

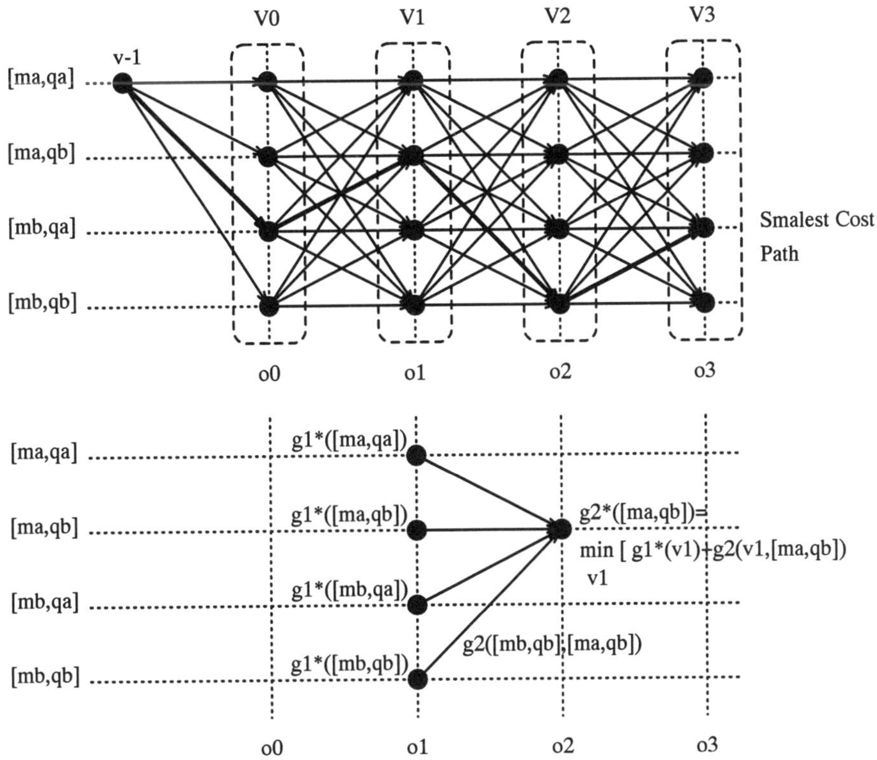

Figure 6.1 The trellis of the lossless MCVC example

to the above example is displayed. The upper trellis in Fig. 6.1 shows the entire trellis for this example whereas the lower trellis shows a specific minimization. The different quantizer and motion vector configurations are indicated on the left and the direction of the scanning path is from left to right starting at region o_0 and ending at region o_3. Each node in the trellis represents a certain decision vector choice for a given region. In the lower trellis, it is shown how $g_2^*([ma, qb])$ is calculated; it can be interpreted in this example as the smallest rate needed to encode region o_0 up to and including region o_2, where region o_2 uses the motion vector ma and the quantizer qb.

In this example we can assign the bit rate needed to encode the DFD of a given region to the associated node, which is the result of using a particular motion vector with a particular quantizer. The transitional bit rate between the nodes occurs because of the DPCM encoding of the DVF. This dependency is the reason why DP is used to solve this example. This bit assignment is not true in general but it is a good motivation for the use of DP to optimize a MCVC. Note that for an exhaustive search $4^4 = 256$ comparisons are necessary, whereas the DP solution requires only $3 * 4^2 + 4 = 52$ comparisons.

6.4 LOSSY MCVC

So far we have considered lossless MCVC. In this section we study the more interesting case of lossy MCVC. Clearly, for a lossy MCVC it does not make sense to minimize the frame rate R_k with no additional constraints, since this would lead to a very high frame distortion D_k.

The most common approach to solve the tradeoff between frame rate and frame distortion is to minimize the total frame distortion D_k subject to a given maximum frame rate R_{max}. Clearly, since the total number of regions, N, is known, minimizing the total distortion is equivalent to minimizing the average distortion. This problem can be formulated in the following way,

$$\min_{v_0,\ldots,v_{N-1}} D_k(v_0,\ldots,v_{N-1}), \qquad \text{subject to:} \quad R_k(v_0,\ldots,v_{N-1}) \leq R_{max}. \tag{6.18}$$

As we have seen in section 4.1, such a constrained discrete optimization problem is generally very hard to solve. When a solution which belongs to the convex hull of the operational rate distortion curve is adequate, then the Lagrangian multiplier method presented in section 3.3 can be employed. We show in section 6.8 that the solutions on the rate distortion curve tend to be quite dense and hence the convex hull approximation is very good. In this application we will use Lagrangian relaxation to relax the constraint so that the relaxed problem can be solved by DP. This is the same strategy employed in [178, 224] for the problem of optimal mode and quantizer selection for H.263.

First we introduce the Lagrangian cost function which is of the following form,

$$J_\lambda(v_0,\ldots,v_{N-1}) = D_k(v_0,\ldots,v_{N-1}) + \lambda * R_k(v_0,\ldots,v_{N-1}), \tag{6.19}$$

where $\lambda \geq 0$ is called the Lagrangian multiplier. We show in section 4.1 that if there is a λ^* such that

$$[v_0^*,\ldots,v_{N-1}^*] = \arg\min_{[v_0,\ldots,v_{N-1}]} J_{\lambda^*}(v_0,\ldots,v_{N-1}) \tag{6.20}$$

leads to $R_k(v_0^*, \ldots, v_{N-1}^*) = R_{max}$, then $v_0^*, \ldots, v_{N-1}^*$ is also an optimal solution to (6.18). It is well known that when λ sweeps from zero to infinity, the solution to problem (6.20) traces out the convex hull of the rate distortion curve, which is a non-increasing function. Therefore bisection [59] could be used to find λ^*. A faster converging algorithm which uses some knowledge about the convexity of the curve is employed in [164] and in section 4.2 we presented an even faster algorithm.

Therefore the first step is to find the optimal solution to problem (6.20). We next show how the original DP approach can be modified to find the global minimum of problem (6.20). For a given λ, let the $g_i(v_{i-a}, \ldots, v_{i+b})$ functions be defined as follows,

$$g_i(v_{i-a}, \ldots, v_{i+b}) = r_i(v_{i-a}, \ldots, v_{i+b}) + \lambda * d_i(v_{i-a}, \ldots, v_{i+b}), \tag{6.21}$$

which implies that, $G_k(v_1, \ldots, v_N) = J_\lambda(v_1, \ldots, v_N)$. This is true since the frame rate and the frame distortion are additive with respect to the region rates and region distortions, which is expressed in assumptions (6.1) and (6.2). Hence the DP algorithm presented in section 6.3 leads to the optimal solution of problem (6.20).

Note that the dual problem, which can be stated as follows,

$$\min_{v_0, \ldots, v_{N-1}} R_k(v_0, \ldots, v_{N-1}), \qquad \text{subject to:} \quad D_k(v_0, \ldots, v_{N-1}) \le D_{max}, \tag{6.22}$$

can be solved with exactly the same technique using the following relabeling of function names,

$$R_k(v_0, \ldots, v_{N-1}) \leftarrow D_k(v_0, \ldots, v_{N-1}), \tag{6.23}$$

$$D_k(v_0, \ldots, v_{N-1}) \leftarrow R_k(v_0, \ldots, v_{N-1}). \tag{6.24}$$

6.5 THE MINIMUM MAXIMUM DISTORTION APPROACH

In this section we show that the DP solution for the lossless case can be used to find the optimal solution for the minimum maximum distortion criterion. As we pointed out in chapter 4, the minimum total distortion approach and the minimum maximum distortion approach are two powerful paradigms for formulating the tradeoff between the rate and the resulting distortion. In chapter

4, we derived two different algorithms which find the optimal solutions for both paradigms in the case of arbitrary quantizer dependencies. In the previous section we used the Lagrangian formulation to find the optimal solution for the minimum total distortion approach. In this section we show how the DP solution for the lossless case can be modified so that the minimum rate case solution can be found for the minimum maximum distortion criterion. We showed in section 4.3 that if we can find the optimal solution for the minimum rate case, then the optimal solution for the minimum distortion case can also be found, using a bisection search. The minimum rate case can be formulated as follows,

$$\min_{v_0,\ldots,v_{N-1}} R_k(v_0,\ldots,v_{N-1}) \quad \text{subject to:} \quad D_k(v_0,\ldots,v_{N-1}) \leq D_{max}, \tag{6.25}$$

where the frame distortion is the maximum of the region distortions,

$$D_k(v_0,\ldots,v_{N-1}) = \max_{i\in[0,\ldots,N-1]} d_i(v_{i-a},\ldots,v_{i+b}). \tag{6.26}$$

The modification of the DP solution for the lossless case follows the basic idea presented in section 4.3, where the transition cost $g_i(v_{i-a},\ldots,v_{i+b})$ is set to infinity when the region distortion exceeds D_{max} and is otherwise set to the region rate. That is

$$g_i(v_{i-a},\ldots,v_{i+b}) = \begin{cases} \infty & : \; d_i(v_{i-a},\ldots,v_{i+b}) > D_{max} \\ r_i(v_{i-a},\ldots,v_{i+b}) & : \; d_i(v_{i-a},\ldots,v_{i+b}) \leq D_{max} \end{cases}. \tag{6.27}$$

The idea behind the above re-definition of the transition cost is that the DP solution will find the sequence of decision vectors, such that the rate is as small as possible, without violating the maximum distortion criterion. If the minimization results in an objective value of infinity, then there was no admissible decision vector sequence which can satisfy the maximum distortion requirement.

6.6 A VIDEO COMPRESSION SCHEME WITH OPTIMAL BIT ALLOCATION BETWEEN DVF AND DFD

In this section we present an example of how the proposed theory can be applied to existing coders and how the understanding of the theory can foster new coders well suited for the proposed optimization algorithm. We will apply the presented theory to the optimal allocation of the bits between the DFD and the DVF for a video coder which is largely based on the ITU standard for very low bit rate video coding H.263 [46].

The presented theory can be used to optimize H.263 with all its options activated. For simplicity, we base our coder on H.263 without the recently added options, which are the "Unrestricted Motion Vector Mode", the "Syntax-based Arithmetic Coding mode", the "Advanced Prediction mode" and the "PB-frames mode". In fact, the proposed coder is almost identical to test model 4 (TMN4) [47] of H.263 with some noteworthy exceptions which we will point out later. The changes we incorporated are mainly to reduce the complexity of the optimization procedure and are meant to lead to a video coder which is ideally suited for the presented theory of optimal bit allocation among DFD and DVF. Note that the proposed coder is closely related to the one we presented in [184].

Because of is popularity, we use the peak signal to noise ratio (PSNR) as the distortion measure. It is defined in the following way,

$$PSNR = 10 * \log_{10} \frac{255^2}{\mathrm{MSE}(\hat{f}_k, f_k)}, \tag{6.28}$$

where $\mathrm{MSE}(\hat{f}_k, f_k)$ is the mean squared error between the luminance channel of the reconstructed and the original frames. In the case of TMN4, the MSE frame distortion can be written as,

$$D_k(v_0, \ldots, v_{N-1}) = \sum_{i=0}^{N-1} d_i(v_i), \tag{6.29}$$

since the region distortion $d_i(v_i)$ only depends on the selected motion vector and the selected quantizer for that region. We use "quarter common intermediate format" (QCIF) sequences, which have dimensions 176×144 pixels. Since TMN4 breaks the frame into 11×9 macro blocks of size 16×16, the regions $o_0, \ldots, o_{N-1}$ are defined to be these macro blocks (see Fig. 6.2). Clearly, the total number of regions, N, equals 99 for this implementation. Note that the TMN4 scanning path is a raster-scan. In other words, the upper left block is encoded first and the lower right block is encoded last.

The frame rate for TMN4 can be written as follows,

$$R_k(v_0, \ldots, v_{N-1}) = \sum_{i=0}^{N-1} r_i(v_{i-11}, \ldots, v_i). \tag{6.30}$$

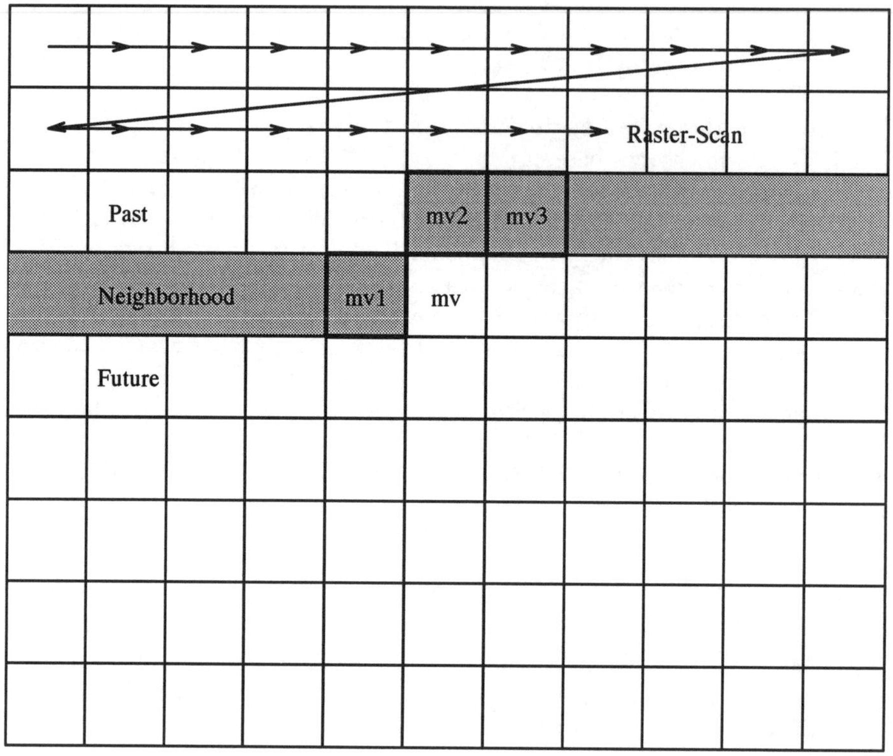

Figure 6.2 The neighborhood needed for TMN4

The reason for $a = 11$ can be found in the way TMN4 encodes the current motion vector. TMN4 uses the vector median of three neighboring motion vectors as the prediction for the current motion vector (see Fig. 6.2). The three motion vectors employed for the prediction belong to the macro block to the left of the current macro block, to the macro block above the current macro block and to the macro block to the right and above the current macro block. The macro block directly above the current macro block has been visited by the raster scan 11 macro blocks earlier than the current macro bock. Therefore, the neighborhood has to contain the information about the encoding decisions made for the last 11 macro blocks, because this is the knowledge needed to make the future of the optimization process independent from its past. Since the computational complexity of the proposed optimal bit allocation algorithm is exponential in a, we would like to keep a as small as possible. The smallest

a, which is still useful when $b = 0$ as in this case, is $a = 1$. In other words, we will employ a single predictor DPCM for the DVF encoding. This is the same way all the other standards encode the DVF. There are other dependencies in TMN4, which we will discuss later, which all can be captured by an a of one. Hence the total frame rate can now be expressed by,

$$R_k(v_0, \ldots, v_{N-1}) = \sum_{i=0}^{N-1} r_i(v_{i-1}, v_i), \tag{6.31}$$

where

$$r_i(v_{i-1}, v_i) = r_i^{QDFD}(v_i) + r_i^{DVF}(v_{i-1}, v_i), \tag{6.32}$$

$r_i^{QDFD}(v_i)$ are the bits needed to encode the DFD of block o_i using the quantizer q_i and the motion vector m_i, and $r_i^{DVF}(v_{i-1}, v_i)$ the bits needed to encode the motion vector difference $(m_i - m_{i-1})$.

In the next two paragraphs we define what we exactly mean by q_i and m_i. Let $e_i \in E_i$ be the encoding mode of macro block o_i, where E_i = {Intra, Inter, Skip, Prediction}. The encoding mode can be set differently for each macro block. When the Intra mode is selected, then the macro block is encoded using a "JPEG"-like scheme and its associated motion vector is set to zero. In the Inter mode, the motion vector is used to create the predicted block and then the difference between the original and the predicted block is encoded using a similar scheme as in the Intra mode. The Skip mode means that the current block is replaced by the block at the same location in the previous reconstructed frame and its motion vector is considered to be zero. TMN4 has all these three modes but the next mode, the Prediction mode, has been introduced by us. This mode is identical to the Inter mode with the exception that no difference signal is sent. Hence when the prediction is good enough, the Prediction mode should be used.

Let $QP_i \in Z_i$ be the DCT domain quantizers for block o_i, where Z_i is the set of all admissible DCT domain quantizers for block o_i. In TMN4, 31 different DCT domain quantizers are admissible. Note the distinction made between quantizers and DCT domain quantizers. The reason for this is that the modes can be considered quantizers too. Therefore we can define the new set of quantizers for block o_i as $q_i = [e_i, QP_i] \in Q_i$ where $Q_i = E_i \times Z_i$. As defined before, $m_i \in M_i$ is the motion vector for block o_i, where M_i is the set of all admissible motion vectors for block o_i.

It is well known that the DC values of the luminance (Y) and chrominance channels (Cb, Cr) of neighboring blocks are highly correlated and therefore an encoding scheme should take advantage of this. One way of exploiting this fact is by encoding the DC values of consecutive Intra blocks by a single predictor DPCM. Therefore an additional dependency has been introduced and $r_i(v_{i-1}, v_i)$ in Eq. (6.32) is now equal to

$$r_i(v_{i-1}, v_i) = r_i^{QDFD}(v_i) + r_i^{DVF}(v_{i-1}, v_i) + r_i^{DC}(v_{i-1}, v_i), \tag{6.33}$$

where $r_i^{DC}(v_{i-1}, v_i)$ is zero whenever both blocks o_{i-1} and o_i are not Intra coded, and equal to the number of bits needed to encode the difference between the DC coefficients of the two Intra coded blocks, otherwise.

For an Intra frame, the DPCM of the DC values of a block is very important since many bits can be saved this way. Hence by using this DPCM the proposed coder will also efficiently encode an Intra frame, such as the first frame of a sequence or the first frame after a scene change.

In TMN4 a quantizer is selected by transmitting a quantizer step size QP. QP is encoded using a modified delta modulation with a range of ± 2. Hence the quantizer step size of block o_i, QP_i, is equal to $QP_{i-1} + \delta_i$, where $\delta_i \in [-2, -1, 0, 1, 2]$. At the beginning of the frame, QP_0 of the first block is coded using PCM. Clearly, this delta modulation introduces another dependency which can be captured by modifying $r_i(v_{i-1}, v_i)$ in Eq. (6.33),

$$r_i(v_{i-1}, v_i) = r_i^{QDFD}(v_i) + r_i^{DVF}(v_{i-1}, v_i) + r_i^{DC}(v_{i-1}, v_i) + r_i^{QP}(v_{i-1}, v_i), \tag{6.34}$$

where $r_i^{QP}(v_{i-1}, v_i)$ represents the bits needed to encode $\delta_i \in [-2, -1, 0, 1, 2]$. It is set to infinity for a QP_i which is out of reach. This will force the optimal path to select only accessible quantizers.

In most MCVCs the blocks are processed along a simple raster scan. Clearly DPCM is most effective when the data is highly correlated. In the presented approach, the motion vector, the DC values (for Intra blocks) and the quantizer step size of the previous block are used as predictors for the values of the current block. Therefore, the higher the correlation between the blocks along the scanning path is, the better the performance of the coding scheme.

Hilbert curves have been used in image and video processing as scanning paths on the pixel level in the luminance domain for lossless coding [159] and lossy coding [119]. They have also been used as a scanning path for the coefficients in the transform domain [39]. In all these cases, the fact that a scanned image according to a Hilbert curve creates an one-dimensional representation of the image, which is more correlated than a raster scan, is exploited.

Since in the presented approach the correlation of the blocks along the scanning path is of foremost importance, a Hilbert curve should be employed for the scanning of the blocks. Since the proposed video coder works with QCIF video sequences, a modified version of a Hilbert scan is used (see Fig. 4.23) since a perfect Hilbert scan requires an image format of $2^n \times 2^n$, where n is an integer. Note that a Hilbert scan is also ideal for higher order predictors since the previous blocks along the scanning path are closer to the current block than in a raster scan.

6.7 IMPLEMENTATION ISSUES

In this section we discuss how the computational complexity can be further reduced by restricting the set of admissible decision vectors and using a fast evaluation of the operational rate distortion functions.

From a theoretical point of view, every possible motion vector of block o_i should be included in the set of admissible motion vectors M_i. This means that in the case of the TMN4 implementation, where the search window is ± 15.5 pixels and the accuracy of the motion estimation is 1/2 pixel, $|M_i| = 63 * 63 = 3969$, which is quite large. Most of these motion vectors are not likely candidates for the optimal path since they do not correspond well to the real motion in the scene and therefore they lead to a high distortion and a high rate.

To make the optimization process faster, such prior knowledge should be used. Even though this is complicated in general, it can be achieved easily in the presented framework by constraining the set M_i of admissible motion vectors of block o_i. We propose the following strategy to constrain the set M_i. An initial motion vector is first found by using block matching with integer accuracy and the sum of absolute error matching criterion. The set M_i is then defined as the set which contains this motion vector plus the K neighboring motion vectors at half pixel locations. This leads to $|M_i| = K + 1$ which can be much smaller than the original size of 3969. Our experiments have shown (see section 6.10) that a $K = 8$ results in a performance loss which is negligible compared to the achieved reduction in computational complexity.

A similar situation arises for the quantizer selection. In TMN4, the quantizer parameter QP_i can take on values between 1 and 31. Since a nearly constant distortion is usually targeted, a reduced admissible quantizer set, which is centered around the quantizer step size which leads to the desired distortion, can be used without any noticeable loss of performance. The set employed in the presented experiments is $Z_i = \{8, 9, 10, 11, 12\}$.

Since for TMN4 (and for most other coders), the Skip and Intra modes imply a zero motion vector, this knowledge can be used to further reduce the set of admissible decision vectors V_i. For the Inter and Prediction encoding modes, $|M_i|$ motion vectors are considered, but for the Skip and Intra encoding modes only the zero motion vector is needed. This leads to the final admissible decision vector set $V_i = (\{\text{Inter,Prediction}\} \times Z_i \times M_i) \cup (\{\text{Skip,Intra}\} \times Z_i \times \{\vec{0}\})$. The cardinality of this set equals $2 * |Z_i| * (|M_i| + 1)$ (recall that the time complexity of DP is $O(N * |V_i|^{a+b+1})$).

Note that we restricted the sets M_i and Z_i to reflect our prior knowledge about the solution. One of the great advantages of DP is that, besides finding a global optimum, it can incorporate difficult constraint sets such as the ones we formulated above.

For every member v_i of the set V_i a new rate and distortion pair needs to be calculated. The number of required DCTs per block is equal to $|M_i| + 1$, since the DFD of every admissible motion vector needs to be transformed and for the Intra encoding mode, the DCT of the original block has to be calculated. In general it takes as many inverse DCTs as it takes DCTs, since for the calculation of the distortion, the reconstructed block must be available. By selecting the mean squared error or a weighted MSE distortion measure, however, these inverse DCTs are not necessary. This is true since the DCT used (DCT-II, [166]), is a distance preserving transformation. This means that the squared sum of the elements in the original domain, is equal to the squared sum of the elements in the DCT domain. Therefore, the mean squared error can be computed in the DCT domain and no inverse DCT operation is required [178].

Using the above admissible decision vector reduction and the fast distortion calculation, the current implementation of the proposed coder requires about three times the amount of time to encode a video sequence than the available TMN4 implementation, on which the proposed coder implementation is based.

6.8 EXPERIMENTS

In this section, the results of the proposed coder which is based on our general theory for optimal bit allocation between DFD and DVF, are compared to

TMN4. Because of the popularity of the minimum total distortion criterion, we use the approach presented in section 6.4, which is based on the Lagrangian multiplier method, to find the bit allocation between DFD and DVF. As we pointed out in section 6.4 the Lagrangian relaxation can only find solutions which belong to the convex hull of the rate-distortion curve. If these solutions are dense enough, the convex hull approximation can for practical purposes be considered the optimal solution. Based on our experiments we have found that the convex hull solution is usually within ±2.5% of the desired solution, which is certainly sufficient for practical systems.

Note that the presented coder, like TMN4, writes a bit stream which is uniquely decodable by our decoder. Hence the listed bit rates are the effective number of bits used and not an estimate of the entropy. Since TMN4 was selected to implement the proposed optimal bit allocation scheme, most parts of the proposed coder and TMN4 are identical. The deviations of the proposed coder from the TMN4 implementation consist of the use of the following:

- First oder DPCM encoding of the DVF.
- Modified Hilbert scan.
- DPCM encoding of the DC values (Y, Cb and Cr) for consecutive intra blocks.
- Optimal bit allocation between DFD and DVF.

The first order DPCM encoding of the DVF is selected to keep the computational complexity of the optimization procedure reasonable. As already mentioned, TMN4 uses a more sophisticated DVF DPCM encoding which involves three previous motion vectors. The Hilbert scanning path is used to maximize the correlation along the scanning path, which improves the efficiency of the first order DPCM. The DPCM encoding of the DC values for consecutive blocks enables the proposed coder to encode Intra frames, such as the first frame, in an efficient and optimal way without having to treat these frames differently.

The first three changes to TMN4 listed above have been incorporated so that the resulting coder presents a good framework for the employed optimal decision strategy. Note that these changes alone do not lead to a better coder than TMN4. As a matter of fact, the use of a first order DPCM encoding of the DVF might reduce the performance of the algorithm. Therefore, all the improvements in the results presented in the following are achieved by the proposed optimal bit allocation between DFD and DVF.

In order to compare TMN4 and the proposed coder, TMN4 was used to encode every 4th frame of the first 200 frames of the QCIF color sequence "Mother and Daughter" with a fixed quantizer step size $QP = 10$. The first frame was Intra coded using the same quantizer step size. Since the "Mother and Daughter" sequence is considered to be recorded at 30 frames/second, this leads to an encoded rate of 7.5 frames/second. The resulting frame rate and frame distortion were used for the comparison between TMN4 and the proposed coder. Remember that the term "frame rate" is used for the number of bits required to encode a certain frame and not for the number of encoded frames per second. The employed distortion measure is the peak signal to noise ratio (PSNR) which is defined in Eq. (6.28).

6.8.1 Matched distortion

The goal of this experiment is to compare the proposed coder with TMN4 when their frame distortions are matched. This can be achieved by setting D_{max}, the maximum frame distortion in Eq. (6.22), equal to the frame distortion of TMN4. Clearly D_{max} changes from frame to frame, following the distortion profile generated by the TMN4 run. The proposed coder will lead to the smallest number of bits needed to encode a given frame for the given maximum distortion D_{max}.

The resulting rate and distortion per frame are displayed in Figs. 6.3, 6.4, 6.5, and 6.6. The average rate and distortion for the entire sequence are listed in Table 6.1 (second row). Note that the distortion of the proposed coder follows the TMN4 distortion extremely closely. Clearly the proposed coder is superior to TMN4 when their frame distortions are matched, based on the resulting difference (3.4 kbits/s on the average) in bit rates.

Fig. 6.7 shows the reconstructed 12th frame of the sequence which is used to predict the 16th frame. The optimal encoding mode selection, the optimal quantizer selection and the optimal motion vector field are displayed in Figs. 6.8, 6.9 and 6.10 for the 16th frame of the "Mother and Daughter" sequence. Note in Fig. 6.8 how the new object (the hand) and the uncovered areas (left of the hand) are Intra coded and the stationary background is replaced by the blocks from the previously decoded frame (Skip mode). Also note the smoothness of the motion vector field in Fig. 6.10, which can be encoded very efficiently by DPCM.

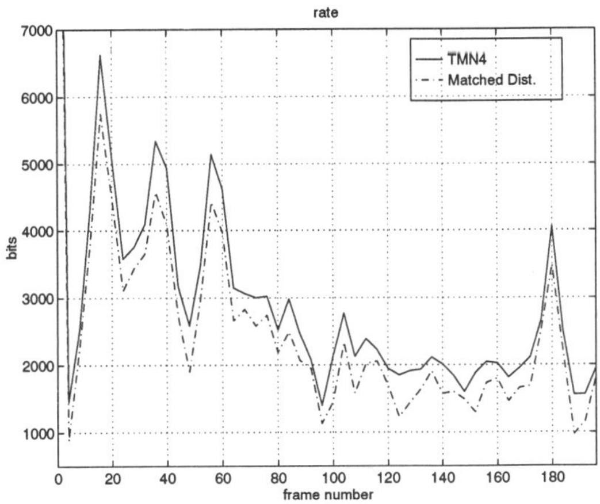

Figure 6.3 Rate comparison between TMN4 and the proposed coder, where the TMN4 distortion is the target distortion of the proposed coder.

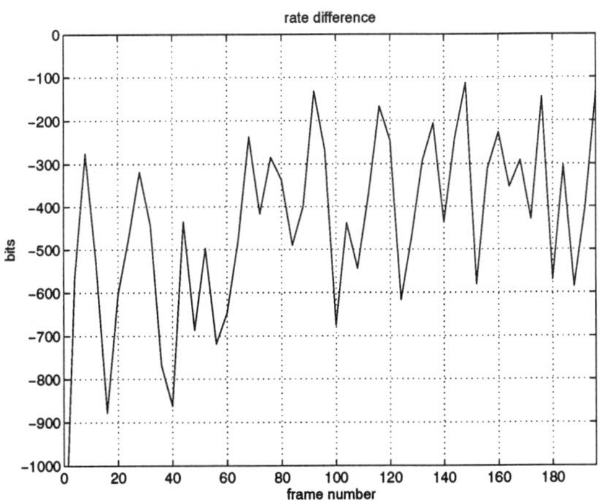

Figure 6.4 Rate difference between TMN4 and the proposed coder, where the TMN4 distortion is the target distortion of the proposed coder.

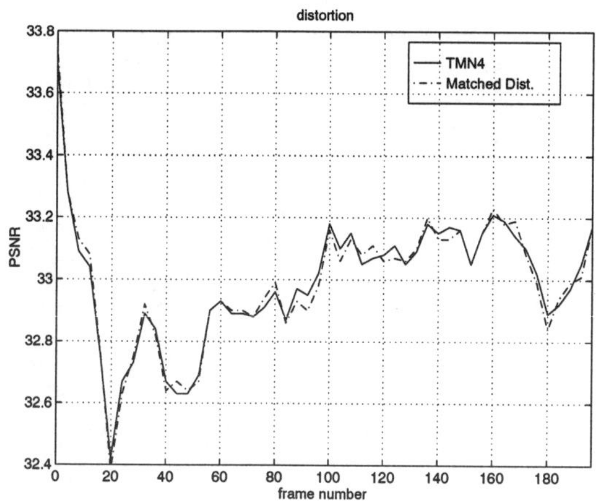

Figure 6.5 Distortion comparison between TMN4 and the proposed coder, where the TMN4 distortion is the target distortion of the proposed coder.

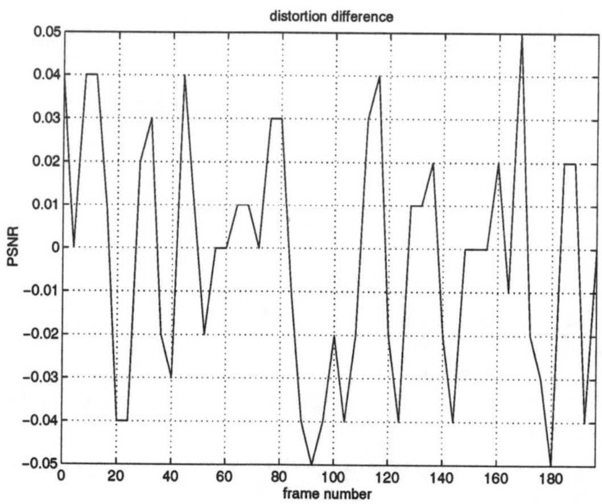

Figure 6.6 Distortion difference between TMN4 and the proposed coder, where the TMN4 distortion is the target distortion of the proposed coder.

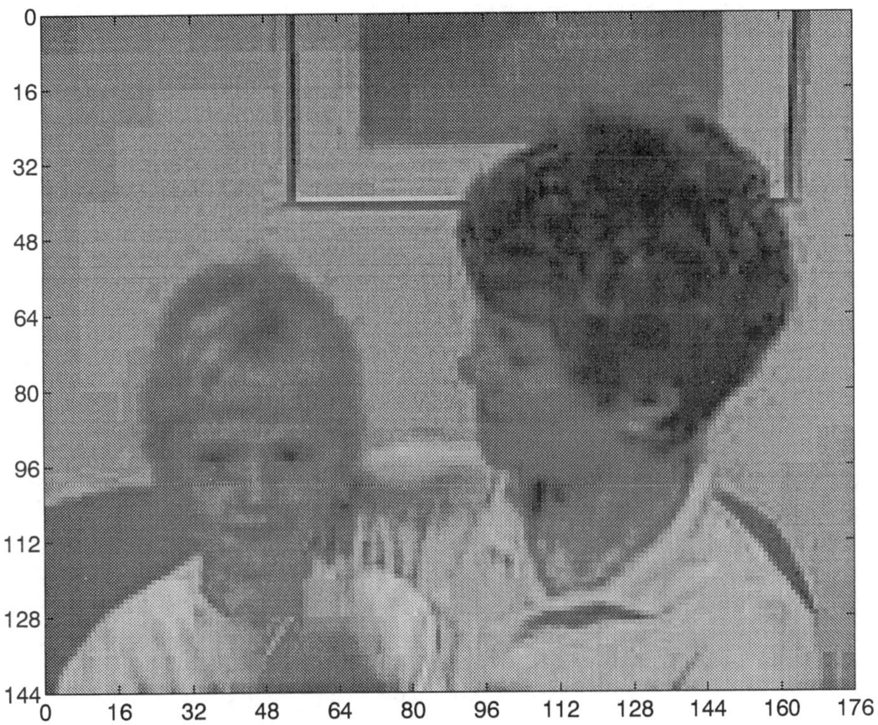

Figure 6.7 The 12th reconstructed frame of the "Mother and Daughter" sequence. This frame is used to predict the 16th frame.

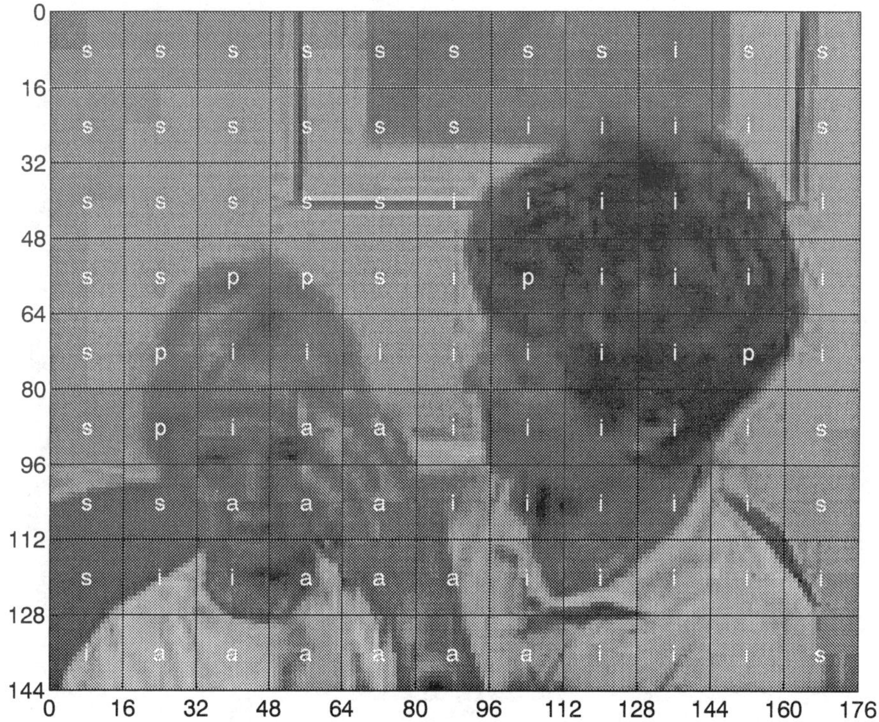

Figure 6.8 The optimal mode selection for the 16th frame of the "Mother and Daughter" sequence. (i) Inter mode, (s) Skip mode, (p) Prediction mode and (a) Intra mode.

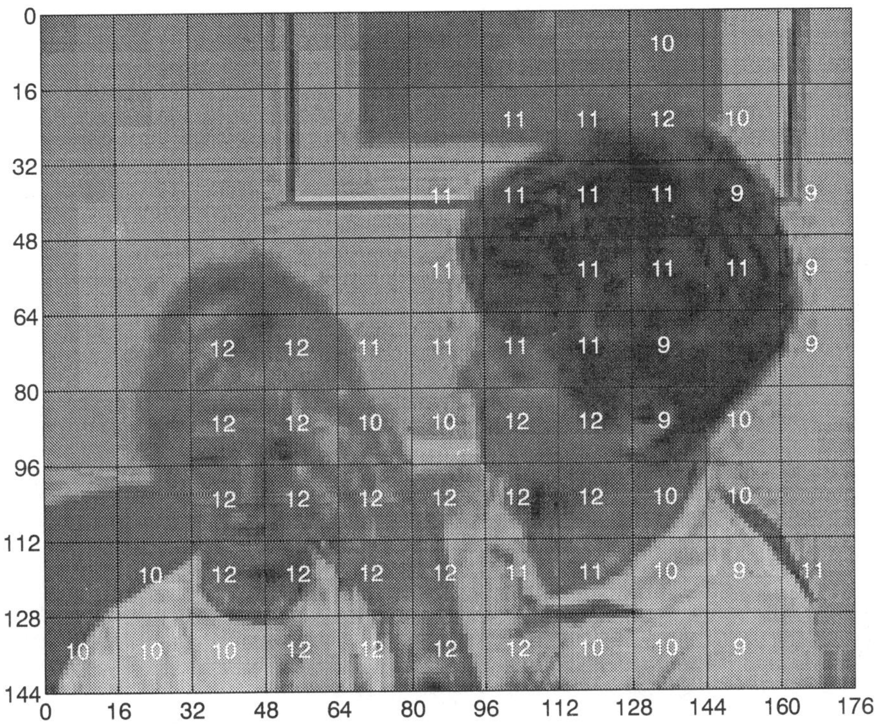

Figure 6.9 The optimal quantizer selection for the 16th frame of the "Mother and Daughter" sequence. The numbers stand for the quantizer step size QP used for that block.

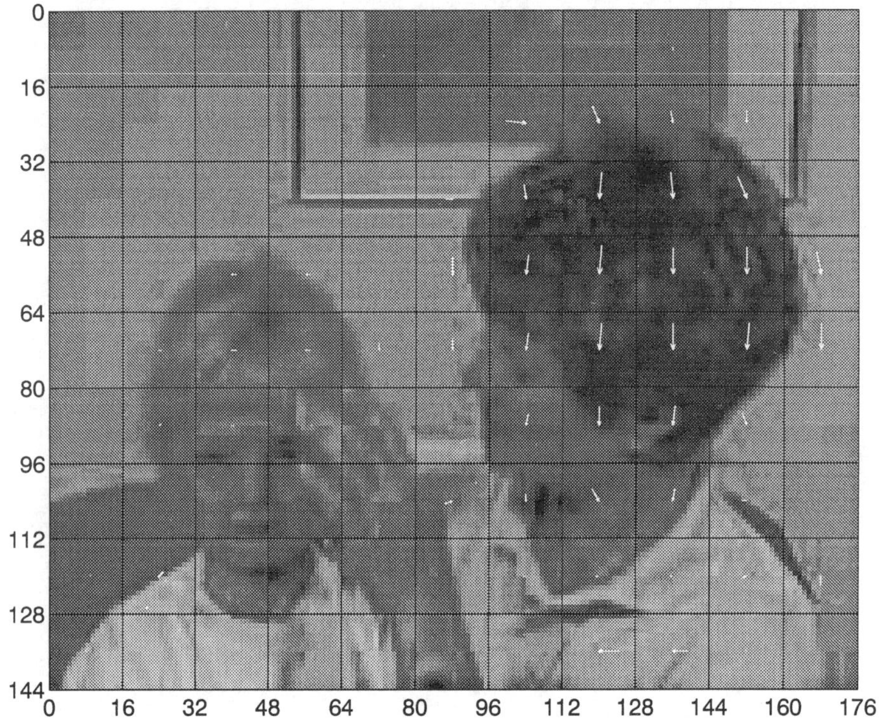

Figure 6.10 The optimal motion vector field for the 16th frame of the "Mother and Daughter" sequence.

	average rate kbits/s	average distortion (average PSNR) dB
TMN4	23.4	33.0
matched distortion	20.0	33.0
matched rate	23.4	33.5
constant distortion	17.3	32.4

Table 6.1 Average rate distortion comparison for the "Mother and Daughter" sequence between TMN4 and the proposed coder for different modes of operation

The influence of the constrained search space on the rate for the matched distortion case

As mentioned in section 6.7, the optimization process can be accelerated by using prior knowledge about the admissible motion vectors and quantizers. In this section we experimentally compare various solutions with differently constrained search spaces. Table 6.2 is discussed which is a collection of encoding results for the "Mother and Daughter" sequence which all have the same distortion profile as TMN4, resulting in an average distortion of 33.0 dB PSNR. In Table 6.2 the following information is shown: in column one, the available encoding modes; in column two, the admissible motion vectors; in column three, the admissible quantizer step sizes; in column four, the cardinality of V_i; and in column five, the resulting bit rate. The coders are listed in decreasing order of search space constraints. In other words, the top coder has the most constrained search space and the bottom coder has the least constrained search space. As expected the bit rate drops as the search space get less constrained but also the computational complexity rises as the square of the cardinality of V_i. For all the results presented in this paper, the coder with a cardinality of $|V_i| = 100$ is used, since for this cardinality, the tradeoff between speed and performance is very good.

It is interesting to notice that the inclusion of the 8 half-pel neighbors in the motion vector search achieved by far the biggest gain in the bit rate, and that additional inclusion of more motion vectors did not improve the result significantly. Hence the TMN4 generated motion vectors are very close to the

	encoding modes	motion vectors	step sizes	$\|V_i\|$	average rate kbits/s
TMN4	Intra, Inter Skip	TMN4	10	NA	23.36
matched Distortion	Intra, Inter Skip, Pred.	TMN4	10	4	22.89
matched Distortion	Intra, Inter Skip, Pred.	TMN4 + 8 half-pel neighbors	10	20	20.56
matched Distortion	Intra, Inter Skip, Pred.	TMN4 + 8 half-pel neighbors	8,9,10, 11,12	100	20.03
matched Distortion	Intra, Inter Skip, Pred.	TMN4 + 80 half-pel neighbors	8,9,10, 11,12	820	19.80

Table 6.2 Average rate comparison for the "Mother and Daughter" sequence between TMN4 and the distortion matched proposed coder with differently constrained search spaces

optimal motion vectors, but the small error made ($\pm$ 0.5 pels) can lead to a loss in performance of about 10%.

6.8.2 Matched rate

In this experiment, the proposed coder is compared to TMN4 when their frame rates are matched. This can be achieved by setting R_{max}, the maximum frame rate from Eq. (6.18), equal to the frame rate obtained by TMN4. Clearly R_{max} changes from frame to frame, following the rate profile of TMN4 and the proposed coder minimizes the resulting frame distortion for the given frame rate.

The resulting rate and distortion per frame are displayed in Figs. 6.11, 6.12, 6.13 and 6.14. The average rate and distortion for the entire sequence are shown in Table 6.1 (third row). Again, note that the rate of the proposed coder follows the TMN4 rate very closely. Besides being able to outperform TMN4 for matched rates, this experiment also shows the enormous potential of this approach with respect to rate control since the optimal coder can follow an arbitrary bit assignment per frame and produces the smallest possible distortion for any given bit budget.

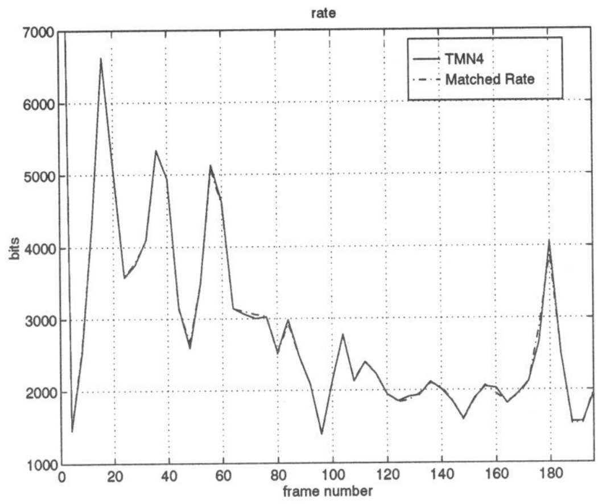

Figure 6.11 Rate comparison between TMN4 and the proposed coder, where the TMN4 rate is the target rate of the proposed coder.

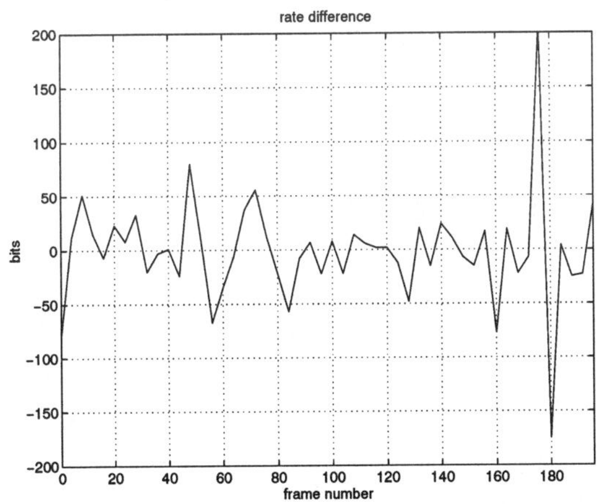

Figure 6.12 Rate difference between TMN4 and the proposed coder, where the TMN4 rate is the target rate of the proposed coder.

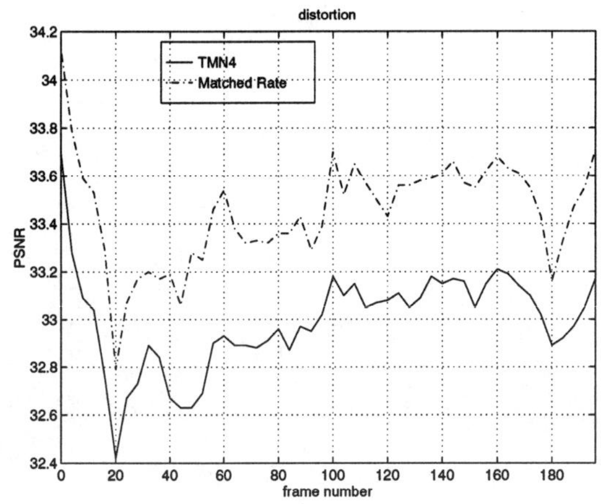

Figure 6.13 Distortion comparison between TMN4 and the proposed coder, where the TMN4 rate is the target rate of the proposed coder.

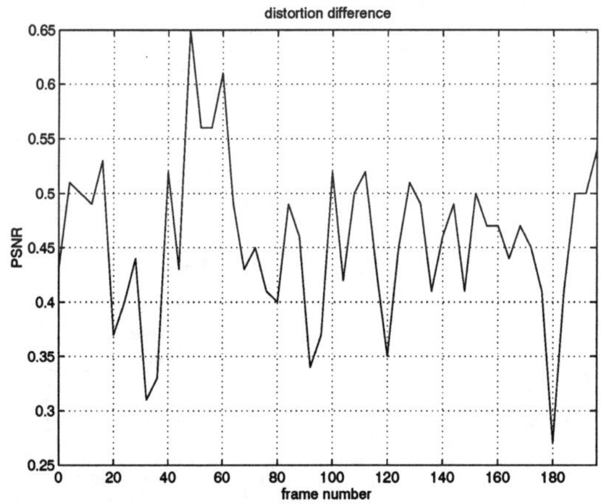

Figure 6.14 Distortion difference between TMN4 and the proposed coder, where the TMN4 rate is the target rate of the proposed coder.

6.8.3 Constant distortion

So far TMN4 and the proposed coder have been compared in terms dictated by the TMN4 run. This gives TMN4 an advantage, since it sets the rate distortion profile, which is then followed by the proposed coder.

An interesting application of the proposed coder is for channels which can accept a variable bit rate, such as an ATM network. For such applications, one would like to keep the distortion constant which can be achieved by setting D_{max} equal to the desired frame distortion. The proposed coder allows therefore for constant quality encoding.

For the experiment discussed next, D_{max} was selected to be equal to the minimum frame PSNR of the TMN4 run, since this is the best quality TMN4 can guarantee over the entire sequence. For this experiment the set of admissible quantizer step sizes Z_i has been changed to $\{9, 10, 11, 12, 13\}$, since on the average, a coarser quantization will be needed to achieve the new target distortion.

The resulting rate and distortion per frame are displayed in Figs. 6.15, 6.16, 6.17 and 6.18. The average rate and distortion for the entire sequence are shown in Table 6.1 (fourth row). Clearly the goal of constant distortion (quality) has been achieved and the resulting average rate is much lower than the TMN4 rate, even though visually these two encoded sequences cannot be distinguished. Some observers even prefer the constant quality sequence over the TMN4 sequence. One possible explanation for this fact is based on the globally optimal selection of the DFD and the DVF. Recall that a Hilbert scan is used for the DPCM encoding of the DVF and a smooth DVF along this path leads to a low bit rate. Hence the optimal solution enforces a global smoothness constraint on the DVF which in turn leads to predicted frames which are more visually pleasing than the ones produced by block matching.

As mentioned before, TMN4 is the test model 4 for the H.263 standard. The most recent test model is TMN6, which is described in [46]. The main changes between TMN4 and TMN6 are the optional modes (described in Annex D, E, F, and G of [46]) which can be switched on in TMN6. These modes are: "Unrestricted Motion Vector mode", "Syntax-based Arithmetic Coding mode", "Advanced Prediction mode", and "PB frame mode". It is interesting to compare our coder with TMN6 in the case when the "Advanced Prediction mode" is switched on, since this leads to a significant improvement over TMN4. This improvement is mostly due to the overlapped block motion compensation and

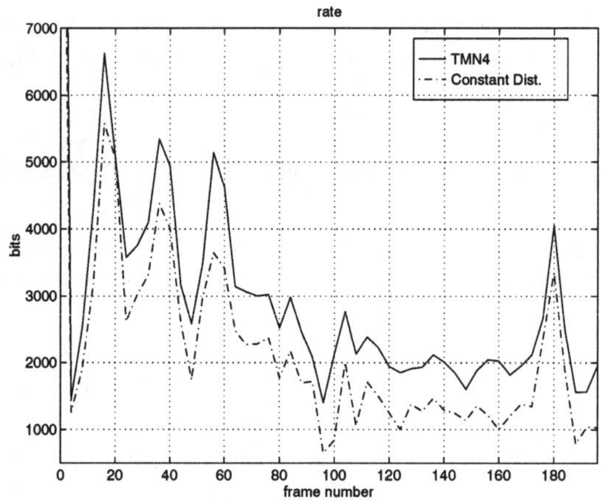

Figure 6.15 Rate comparison between TMN4 and the proposed coder, where the distortion of the proposed coder is fixed.

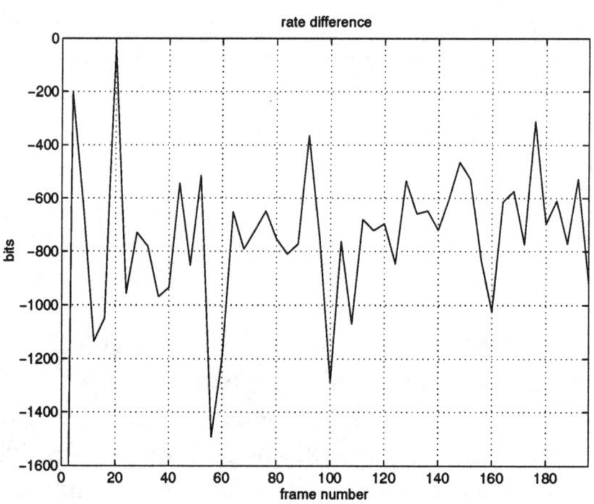

Figure 6.16 Rate difference between TMN4 and the proposed coder, where the distortion of the proposed coder is fixed.

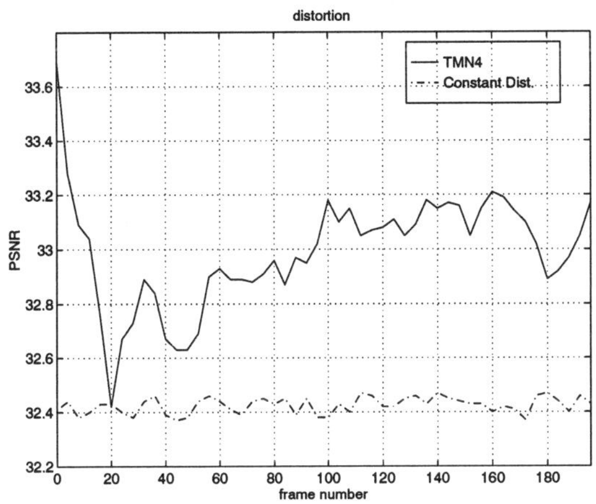

Figure 6.17 Distortion comparison between TMN4 and the proposed coder, where the distortion of the proposed coder is fixed.

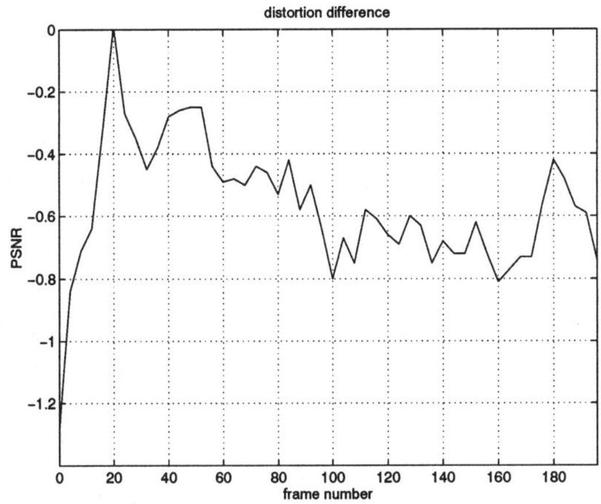

Figure 6.18 Distortion difference between TMN4 and the proposed coder, where the distortion of the proposed coder is fixed.

	average rate kbits/s	average distortion (average PSNR) dB
TMN6	21.0	32.9
matched distortion	19.9	32.9
matched rate	21.0	33.1
constant distortion	17.0	32.3

Table 6.3 Average rate distortion comparison for the "Mother and Daughter" sequence between TMN6 and the proposed coder for different modes of operation

the fact that each of the four luminance blocks in a macro block can have its own motion vector. Both features lead to a better prediction of the current frame and hence to a better compression performance. Note that our scheme does not use these features, but they could be added. Hence to some degree we are comparing "apples with oranges", but nevertheless, as Table 6.3 shows, the proposed scheme still outperforms H.263, even in its most advanced form.

6.9 SUMMARY

In this chapter, we have presented a general theory for the optimal bit allocation between displacement vector field (DVF) and displaced frame difference (DFD). The theory can be applied to any region-based motion compensated video coders (MCVC), which include all current video standards. We first considered a lossless MCVC and derived the optimal bit allocation algorithm which is based on dynamic programming (DP). We then addressed the problem of lossy MCVC and showed that Lagrangian relaxation and DP can find the convex hull approximation to the optimal solution for the minimum total distortion criterion. We also introduced the modifications necessary for DP to find the optimal solution for the minimum maximum distortion criterion.

We next presented a video coder which is largely based on H.263 and uses this optimal bit allocation between the DVF and the DFD. We pointed out the changes we incorporated to reduce the computational complexity and presented results which clearly show the superiority of this coder. We showed that the presented theory can be used to optimize existing video coders. We also showed that it can foster new coders which are designed in such a way that the optimization procedure can be achieved in real time. We pointed out in the matched rate experiments that a video coder employing the optimal bit allocation scheme for the frame encoding is ideal for a rate control algorithm.

One of the main applications of this theory is the evaluation of changes to an existing coding scheme. Conceptually every encoding scheme can be broken down into methods and decisions. The problem in evaluating a certain method is that a decision rule needs to be formulated. During the design process, one can use the presented theory to evaluate the method first (such as a new encoding mode, like the Prediction mode we used), employing the optimal decisions found by the proposed algorithm. If the results are satisfactory, one can use the statistical data of the optimal algorithm to formulate a fast heuristic and evaluate that heuristic versus the optimal decisions. Hence the proposed theory can be used to separately evaluate the method and the decision rule which should speed up the development of new video coders.

7

A VIDEO COMPRESSION SCHEME WITH OPTIMAL BIT ALLOCATION AMONG SEGMENTATION, MOTION AND RESIDUAL ERROR

In this chapter we present a theory for the optimal bit allocation among quad-tree (QT) segmentation, displacement vector field (DVF) and displaced frame difference (DFD) [183, 176]. The theory is applicable to any variable block size motion compensated video coder (VBSMCVC), where the variable block sizes are encoded using the QT structure, the DVF is encoded by first order differential pulse code modulation (DPCM), the DFD is encoded by a block-based scheme and an additive distortion measure is employed.

The presented theory is based on the optimal QT-based motion estimator proposed in section 5.2. For such an estimator, the goal is to find the QT decomposition and the DVF such that the DVF energy is minimized, subject to a total rate constraint, where the total rate is the sum of the DVF rate and the QT encoding rate. In this chapter we extend this scheme to include the optimal selection of the DFD quantizers and hence the goal is now to find the QT decomposition, the DVF and the quantizers so that the distortion between the reconstructed and the original frame is minimized.

7.1 INTRODUCTION

One of the drawbacks of most video coders is the fixed block size used for motion estimation and encoding of the DFD. This is the case with the video coding standards with the only exception represented by H.263 [46], the latest video coding standard, which offers the "Advanced Prediction Mode". This mode contains the option of splitting a 16×16 macro block into four 8×8 blocks. If the macro block is split, then a separate motion vector is transmitted for each of the four 8×8 blocks.

It is well known, that the block size controls the tradeoff between the reliability of the DVF estimate, which increases with the size of the block, and the ability to reduce the energy of the DFD, which decreases with the size of the block [16]. Therefore, a variable block size should be used which adapts to the contents of the scene. Another motivation for the use of a variable block size comes from the object-oriented video compression approach. A fixed block size is equivalent to the claim that every object in the scene is made up of equally sized square blocks, which is certainly not the case. An object-oriented approach finely segments the scene into objects with different motion. There are two major problems with an object-oriented approach: (a) An accurate estimate of the segmentation is required to find the motion information, but on the other hand, one needs the motion information to find the segmentation; (b) The segmentation needs to be encoded and this requires a considerable part of the available bit rate, especially at low bit rates, since the segmentation can be very detailed.

A variable block size coder is a good compromise between the detailed segmentation of the object-oriented approach and the coarse and arbitrary segmentation of a fixed block size coding scheme. It can adapt to a scene which is not possible for a fixed block size approach. The bit rate needed for the transmission of the variable block size is much smaller than the bit rate required for the representation of the segmentation of an object-oriented scheme.

The nesting of blocks of different sizes in the image plane requires some underlying structure. For this purpose, most variable block size motion compensated video coders (VBSMCVC) use the quad-tree (QT) data structure to segment the image plane [96, 200, 195]. A fundamental problem of a VBSMCVC is the bit allocation between the QT, DFD and DVF. In this chapter we present a general theory which uses operational rate distortion curves to solve this problem for a finite set of admissible quantizers and motion vectors.

There have been previous attempts to solve the optimal tradeoff between segmentation, motion and residual error. In [195], a VBSMCVC is presented which uses a mean value QT structure to encode the DFD. The motion estimation is based on block matching and is done for a fixed block size. The optimization of the DVF with respect to the resulting QT is achieved by a local exhaustive search where every motion vector is perturbed by ± 1 pixel and the resulting DFD is encoded using the QT structure. In this scheme, the block size for the motion estimation is kept constant and is not adapted to the scene contents.

This leads to an inefficient encoding of the DFD. Also the tradeoff between the QT, the DFD and the DVF is not optimal in the rate distortion sense.

In [123], the tradeoff between the DVF and the DFD is addressed. This is done in the framework of a multi grid block matching scheme. A stochastic model is assumed for the distribution of the DFD and the entropy of a given block is calculated based on observed statistics. This entropy is then used to decide if a block should be split into four smaller blocks with their own motion vectors, or if the block should be kept as a basic unit. In this scheme, the cost for encoding the segmentation is not taken into account. Also the entropy estimation is based on a stochastic model which assumes that the DFD is directly encoded, i.e., no transform coding or vector quantization is applied. Furthermore the DVF is sent using entropy coding of the motion vectors, but because of the high spatial correlation in the DVF, a DPCM-based scheme should be used. The biggest drawback is that the resulting distortion is not considered in this scheme and therefore any reduction in the bit rate can easily be offset by an increase in distortion.

In [96], a variable block size motion estimator is presented. It is implied that the motion vectors are encoded by pulse code modulation and hence the resulting optimization procedure is quite simple and in fact equivalent to the one presented in [200]. This scheme has the same shortcoming as the previous one. Since the DVF exhibits a large spatial correlation, an efficient video coding scheme should use a predictive approach for encoding the DVF.

This chapter is organized as follows: In section 7.2, we define the problem under consideration. In section 7.3 we derive the optimal solution for lossless VBSMCVC. This solution is then extended in section 7.4 to include lossy VBSMCVC for the minimum total distortion criterion and the minimum maximum distortion criterion. In section 7.5 we develop a lossy video coder based on the presented theory and in section 7.6 we present experimental results with this coder. This chapter is summarized in section 7.7.

7.2 NOTATION AND ASSUMPTIONS

In this section we introduce the necessary notation and state the assumptions which will be used in the rest of the chapter. Our study of the optimal bit allocation between the QT, DVF and DFD is restricted to the frame level. In other words, we do not attempt to optimally allocate the bits among the different frames of a video sequence. The reader interested in this problem is referred to [163]. For the rest of this chapter we assume that a rate control algorithm has set the maximum number of bits available (R_{max}) or the maximum acceptable distortion (D_{max}) for a given frame. In section 4.5 we introduced a general scheme for finding the optimal QT decomposition and the optimal quantizers in the rate distortion sense for a QT with leaf dependencies. We used that general scheme to derive the optimal QT-based motion estimator proposed in section 5.2. In section 5.2 we used the fact that each motion vector can be considered a quantizer operating on the block it belongs to. In this chapter we use the same route, but we keep the distinction between motion vectors and quantizers and re-define the states used in section 5.2. This will still enable us to invoke the optimal solution proposed in section 5.2, without losing track of what is a quantizer and a motion vector in the conventional sense.

We assume that the current frame is segmented by a QT and we use the notation introduced in section 4.4.1. Let $q_{l,i} \in Q_{l,i}$ be the quantizer for block $b_{l,i}$, where $Q_{l,i}$ is the set of all admissible quantizers for block $b_{l,i}$. Let $m_{l,i} \in M_{l,i}$ be the motion vector for block $b_{l,i}$ where $M_{l,i}$ is the set of all admissible motion vectors for block $b_{l,i}$. Let $s_{l,i} = [l, i, q_{l,i}, m_{l,i}] \in S_{i,l} = \{l\} \times \{i\} \times Q_{l,i} \times M_{l,i}$ be the local state for block $b_{l,i}$, where $S_{l,i}$ is the set of all admissible state values for block $b_{l,i}$. Let $x = [l, i, q, m] \in X = \bigcup_{l=N}^{n_0} \bigcup_{i=0}^{4^{N-l}-1} S_{l,i}$ be the global state and X the set of all admissible state values. Finally let $x_0, \ldots, x_{N_{\mathcal{T}}-1}$ be a global state sequence, which represents the left-to-right ordered leaves of a valid QT $\mathcal{T}$.

We assume that the frame distortion $D_k(x_0, \ldots, x_{N_{\mathcal{T}}-1})$ of the reconstructed frame is the sum of the individual block distortions $d(x_j)$. That is,

$$D_k(x_0, \ldots, x_{N_{\mathcal{T}}-1}) = \sum_{j=0}^{N_{\mathcal{T}}-1} d(x_j). \tag{7.1}$$

Most commonly used distortion measures, such as the mean squared error (MSE), a weighted MSE or the peak signal to noise ratio (PSNR) fall into this class.

We assume that the DVF is encoded by first order DPCM. In other words, the difference between consecutive motion vectors is encoded using entropy coding. In chapter 6 we have developed a theory for the optimal bit allocation between DFD and DVF for motion compensated video coders with fixed segmentation, which also covers higher order DPCM schemes. Because the segmentation is not fixed in this chapter, but is encoded as a QT structure, we have to restrict the DPCM to a first order DPCM, so that we are able to formulate an efficient optimization procedure.

Because of the variable block sizes, we add the constraint that blocks which belong to the same parent, need to be encoded in sequence. In section 4.4, we defined optimality for a scanning path, which simply states that consecutive blocks along a scanning path must be spatial neighbors, or in other words, they must share an edge. This definition of optimality is meaningful since we would like to encode the DVF using a first order DPCM along the scanning path. We also introduced a procedure to recursively generate scanning paths for any QT decomposition, given that the scanning path for the completely decomposed QT is known. We derived a necessary and sufficient condition for an optimal scanning path for the completely decomposed QT, which means that the recursive procedure will result in an optimal scanning path for any QT decomposition. In section 4.4.3 we introduced a special scanning path which is based on a Hilbert curve and results in square blocks using the proposed recursive procedure. Note that the theory to be developed next does not require that an optimal scanning path is used but only that blocks with a common parent block are scanned in sequence. Nevertheless, in all the examples in the rest of the chapter, we will use a Hilbert scan.

Based on the first order DPCM assumption, the frame rate $R_k(x_0, \ldots, x_{N_{\mathcal{T}}-1})$ can be expressed as follows,

$$R_k(x_0, \ldots, x_{N_{\mathcal{T}}-1}) = \sum_{j=0}^{N_{\mathcal{T}}-1} r(x_{j-1}, x_j), \tag{7.2}$$

where $r(x_{j-1}, x_j)$ is the block rate which depends on the encoding of the current and the previous block, since the motion vector difference between these two blocks is entropy coded.

7.3 LOSSLESS VBSMCVC

In this section we apply the derivation found in section 4.5.3 to the case of a lossless VBSMCVC. Since the reconstructed frame is identical to the original frame, the frame distortion is equal to zero and the goal is to minimize the required frame rate for QT, DVF and DFD. This can be stated as follows,

$$\min_{x_0, \ldots, x_{N_{\mathcal{T}}-1}} R_k(x_0, \ldots, x_{N_{\mathcal{T}}-1}). \tag{7.3}$$

Since this is a lossless VBSMCVC, the DFD is not quantized, but encoded losslessly. With a slight abuse of notation, let $q_{l,i}$ represent different lossless encoding schemes for block $b_{l,i}$ (i.e., DPCM with different predictor order, etc.), instead of different quantizers.

The solution to this specific problem, i.e., finding the minimum rate for a lossless encoding scheme where the scene is segmented by a QT, has been treated in a general fashion in section 4.5.3 and the results from that section directly apply. In other words, the solution to the lossless VBSMCVC can be found by finding the shortest path through the multilevel trellis presented in section 4.5.4. In this chapter, we will not repeat the algebraic derivation which resulted in the dynamic programming (DP) recursion formula, but we will use the multilevel trellis to give some insight into this specific application of the general algorithm presented in 4.5.

Note that the dependency in the QT decomposition comes from the fact that the leaves of the QT reflect the block sizes used for motion estimation, and that the difference between two consecutive motion vectors (which do not necessarily belong to blocks of the same size) along the scanning path is encoded by a variable length code. Since the code words for small differences between motion vectors are shorter than for large differences, the optimal solution will lead to an inhomogeneous, but smooth motion field, i.e., a motion field which reflects the real motion quite well.

7.3.1 Multilevel trellis

The form of the objective function in Eq. (7.2) of the optimization problem of Eq. (7.3) suggests that DP should be used to find the optimal solution efficiently. To be able to employ forward DP (the Viterbi algorithm), a DP recursion formula needs to be established. A graphical equivalent of the DP recursion formula (for first order neighborhoods, i.e., first order DPCM) is a trellis where the admissible nodes and the permissible transitions are explicitly indicated. Consider Fig. 7.1a which represents the multilevel trellis for a 32×32 image block ($N = 5$), with a QT segmentation developed down to level 3 ($n_0 = 3$, block size 8×8). The QT structure is indicated by the white boxes with the rounded corners. These white boxes are not part of the trellis used for the Viterbi algorithm but indicate the set of admissible state values $S_{l,i}$ for the individual blocks $b_{l,i}$. The black circles inside the white boxes are the nodes of the trellis (i.e., the state values). Note that for simplicity, only two trellis nodes per QT node are indicated, but in general, a QT node can contain any

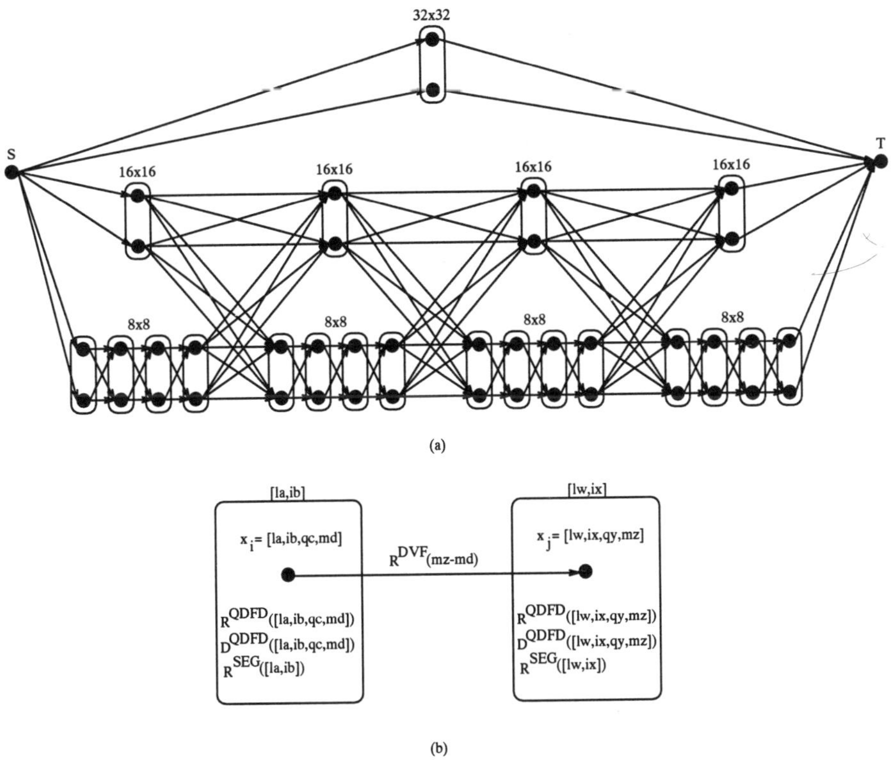

Figure 7.1 The multilevel trellis for $N = 5$ and $n_0 = 3$.

number of trellis nodes. The auxiliary nodes, start and termination (S and T) are used to initialize the DPCM of the motion vectors and to select the path with the smallest cost.

Each of the trellis nodes represents a different way of encoding the block it is associated with, as indicated in Fig. 7.1b. In this figure two nodes $x_i = [la, ib, qc, md]$ and $x_j = [lw, ix, qy, mz]$ of the multi level trellis are shown, where la and lw are the levels to which the nodes belong, ib and ix are the indices of the nodes within their levels, qc and qy are the quantizers associated with the nodes and md and mz are the motion vectors associated with the nodes. In Fig. 7.1b it is also indicated that each of the nodes contains the rate R^{QDFD} (and distortion D^{QDFD}, for the lossy VBSMCVC) occurring by predicting the associated block with the given motion vector and encoding the resulting DFD with the given quantizer. Furthermore the segmentation cost R^{SEG} which only depends on the level and the index of the block, is also contained within the node. The rate for the first order DPCM of the DVF, R^{DVF}, depends on the difference between the two motion vectors mz and md and it is therefore a transition cost between nodes x_i and x_j.

Clearly, finding the shortest path from S to T through the above trellis results in the QT decomposition, the DVF and the quantizer selection which minimizes the total rate, when the length of the path is measured as the sum of the rates along the path. This can be expressed mathematically by defining $r(x_{j-1}, x_j)$ of Eq. (7.2) as follows,

$$r(x_{j-1}, x_j) = R^{SEG}(x_j) + R^{QDFD}(x_j) + R^{DVF}(x_{j-1}, x_j). \tag{7.4}$$

7.4 LOSSY VBSMCVC

So far we have considered lossless VBSMCVC. In this section we study the more interesting case of lossy VBSMCVC. Clearly for a lossy VBSMCVC it does not make sense to minimize the frame rate R_k with no additional constraints, since this would lead to a very high frame distortion D_k.

7.4.1 The minimum total distortion criterion

The most common approach to solve the tradeoff between the frame rate and the frame distortion is to minimize the frame distortion D subject to a given maximum frame rate R_{max}. This problem can be formulated in the following way,

$$\min_{x_0,\ldots,x_{N_T-1}} D(x_0, \ldots, x_{N_T-1}), \quad \text{subject to:} \quad R(x_0, \ldots, x_{N_T-1}) \leq R_{max}. \tag{7.5}$$

In section 3.3 we introduced the Lagrangian multiplier method for discrete optimization problems, and we will now apply this method to find the convex hull solutions of the above problem. First we introduce the Lagrangian cost function which is of the following form,

$$J_\lambda(x_0, \ldots, x_{N_T-1}) = D(x_0, \ldots, x_{N_T-1}) + \lambda * R(x_0, \ldots, x_{N_T-1}), \tag{7.6}$$

where $\lambda \geq 0$ is called the Lagrangian multiplier. We have shown in section 3.3 that if there is a λ^* such that

$$[x_0^*, \ldots, x_{N_T-1}^*] = \arg \min_{x_0, \ldots, x_{N_T-1}} J_{\lambda^*}(x_0, \ldots, x_{N_T-1}) \tag{7.7}$$

leads to $R(x_0^*, \ldots, x_{N_T-1}^*) = R_{max}$, then $x_0^*, \ldots, x_{N_T-1}^*$ is also an optimal solution to the optimization problem stated in Eq. (7.5). It is well known that when λ sweeps from zero to infinity, the solution to problem (7.7) traces out the convex hull of the rate distortion curve, which is a non-increasing function. Hence bisection or the fast convex search we presented in section 4.2 can be used to find λ^*.

Therefore the problem at hand is to find the optimal solution to problem (7.7). We will show how the DP approach presented in the previous section can be modified to find the global minimum of problem (7.7). Again we will use the trellis in Fig. 7.1a to explain the modification, and the algebraic derivation of this result can be found in section 4.5.6, where we discuss the optimal solution for the general QT decomposition problem with leaf dependencies.

In the trellis of Fig. 7.1a, for a given λ, the length of a path is defined as the sum of the block distortions plus the sum of the block rates which are weighted by the Lagrangian multiplier λ. In other words, the total Lagrangian cost function is calculated by the following sum,

$$J_\lambda(x_0, \ldots, x_{N_T-1}) = \sum_{j=0}^{N-1} \{d(x_j) + \lambda * r(x_{j-1}, x_j)\}, \tag{7.8}$$

where the block rate $r(x_{j-1}, x_j) = R^{SEG}(x_j) + R^{QDFD}(x_j) + R^{DVF}(x_{j-1}, x_j)$ and the block distortion equals the distortion of the quantized DFD, $d(x_j) = D^{QDFD}(x_j)$. Using the above definitions, the shortest path identifies the optimal solution of the unconstrained problem. Hence an iterative scheme can be used to find the optimal λ^*, such as the very fast convex search proposed in section 4.2, so that the resulting rate is equal or almost equal to R_{max}.

As with all Lagrangian approaches, the dual problem, which can be stated as follows,

$$\min_{x_0, \ldots, x_{N_T-1}} R(x_0, \ldots, x_{N_T-1}), \quad \text{subject to:} \quad D(x_0, \ldots, x_{N_T-1}) \leq D_{max}, \tag{7.9}$$

can be solved with exactly the same technique using the following relabeling of function names,

$$R(x_0, \ldots, x_{N_T-1}) \leftarrow D(x_0, \ldots, x_{N_T-1}), \tag{7.10}$$

$$D(x_0, \ldots, x_{N_{\mathcal{T}}-1}) \quad \leftarrow \quad R(x_0, \ldots, x_{N_{\mathcal{T}}-1}). \tag{7.11}$$

7.4.2 The minimum maximum distortion criterion

So far we have presented a scheme which results in the optimal QT decomposition, optimal DVF and optimal quantizer selection when the objective is to minimize the total distortion. We now introduce the adjustments necessary to the proposed approach, so that the resulting scheme is optimal with the minimum maximum distortion criterion. As we have shown in section 4.3, if we can solve the minimum rate problem optimally, the minimum distortion problem can also be solved optimally using an iterative approach. Hence we will only discuss the minimum rate case, which can be formulated as follows,

$$\min_{x_0,\ldots,x_{N_{\mathcal{T}}-1}} R(x_0, \ldots, x_{N_{\mathcal{T}}-1}), \quad \text{subject to:} \quad D(x_0, \ldots, x_{N_{\mathcal{T}}-1}) \leq D_{max}, \tag{7.12}$$

where

$$D(x_0, \ldots, x_{N_{\mathcal{T}}-1}) = \max_{j\in[0,\ldots,N_{\mathcal{T}}-1]} \{d(x_j)\}. \tag{7.13}$$

Since the QT decomposition results in blocks of different sizes, it is important to realize how the block distortions $d(x_j)$ should be defined so that the desired effect of a best "worst case" is achieved. We propose to define the block distortion $d_{l,i}(q,m)$ (recall that $x = [l,i,q,m]$, hence $d(x_j)$ is equivalent to $d_{l,i}(q,m)$) of block $b_{l,i}$ which is encode using the motion vector m and the quantizer q, in the following recursive fashion, where $d_{n_0,i}(q,m)$ is the MSE (or signal to noise ratio, SNR) of block $b_{n_0,i}$,

$$d_{l,i}(q,m) = \max_{k\in[0,\ldots,3]} \{d_{l-1,4*i+k}(q,m)\}. \tag{7.14}$$

Therefore the MSE of every block at level n_0 is guaranteed to be below or equal to D_{max}, and this is achieved with the fewest bits possible.

Again we will use the trellis of Fig. 7.1a to explain the necessary changes in the algorithm, and an algebraic derivation of this concept can be found in section 4.5.8. The basic idea behind the minimum rate approach is to set the transition cost to go from node x_{j-1} to node x_j to infinity when the block distortion $d(x_j) = D^{QDFD}(x_j)$ is greater than the maximum distortion D_{max}. Setting a transition cost to infinity is the same as removing the corresponding directed edge (x_{j-1}, x_j) from the trellis in Fig. 7.1a. After removing all the edges which arrive at nodes which would result in a block distortion greater than D_{max}, all block distortions in the shortest path must be below or equal to the maximum distortion. Since the goal is to find a path with the smallest rate which does not violate the maximum block distortion criterion, the length of the path is simply measured by the resulting bit rate. Clearly the shortest path will result in the QT decomposition, the DVF and the quantizer selection with the smallest bit rate which does not violate the maximum distortion condition. In other words, the minimum rate approach for a minimum maximum distortion D_{max} is identical to the lossless minimum rate approach of section 7.3, given that all edges which end in a node with a block distortion greater than D_{max} are removed from the trellis in Fig. 7.1a.

As pointed out above and discussed in section 4.3, the minimum distorting problem for the minimum maximum distortion criterion can now be solved by iteratively solving the minimum rate problem as discussed in this section.

7.5 IMPLEMENTATION

So far the general theory for a VBSMCVC has been discussed. In this section, a video coder is presented which is based on the above introduced theory. Because of the popularity of the minimum total distortion criterion, the presented coder is based on the approach introduced in section 7.4, which finds the optimal bit allocation for this criterion. This coder is henceforth called the optimal coder. In this scheme, the lossy encoding of the DFD is accomplished in the discrete cosine transform (DCT) domain, which is the most common way of encoding the DFD. In fact the DCT, quantization, run length and entropy coding are exactly the same as in test model number four, (TMN4 [47]), for the H.263 standard.

In TMN4, the quantizer step size QP can have 31 different values but for the presented results QP was fixed to be equal to 10. The only exception to this is the "constant distortion" scheme presented later, where QP is set to 11. In the presented implementation, each block has an encoding mode which can be Skip, Prediction, Inter or Intra. If the encoding mode is set to Skip, then the block is just replaced by the block at the same location in the previously reconstructed frame. In the case of the Prediction encoding mode, the block is replaced by a motion compensated block from the previously reconstructed frame but still, no DFD is sent. The Inter mode differs from the Prediction mode in that the lossy encoded DFD is also sent. If the Intra mode is selected, then no prediction is used and the block is just Intra encoded, i.e., the original image block is encoded. The quantizer step size $QP = 10$ and the encoding modes of a given block lead to the set of all admissible quantizers, $Q_{l,i} = \{\text{Skip,Prediction,Inter,Intra}\} \times \{10\}$.

The rate distortion points of the Intra and Inter nodes at level $n_0 = 3$ are calculated in the DCT domain [178]. For bigger block sizes, the rate distortion points are calculated by using the rate distortion points of their child nodes. This can be achieved since in the presented implementation, blocks greater than 8×8 which are Inter or Intra coded are encoded by coding their 8×8 child nodes.

In the presented examples, the optimal coder works with video sequences which are in the quarter common intermediate format (QCIF), which is of dimension 176×144 pixels. The theory has been developed for images of size $2^n \times 2^n$, where n is an integer, but it can be easily adapted to another image format such as QCIF.

The smallest block size in the employed QT decomposition is 8×8 ($n_0 = 3$) and the largest block size is 256×256 ($N = 8$). Since the QCIF format is 176×144, the 256×256 QT has been pruned such that only nodes which correspond to an image region are considered. Hence some of the nodes have fewer than four children and this is taken into account when the segmentation cost is distributed recursively. The smallest block size is set to 8×8 since the efficiency of the run length coding drops significantly for 4×4 blocks. Clearly a vector quantization based scheme could easily include 4×4 blocks or even smaller ones but for this implementation, only a DCT-based encoding of the DFD has been considered which will lead to a fair comparison in section 7.6, since the comparison coder and the optimal coder work with the same DFD encoding technology.

The optimal scanning path has been developed for an image of size $2^n \times 2^n$ where n is an integer. Since the proposed coder is applied to QCIF sequences (176×144), only the 128×128 block in the lower left corner is scanned by a pure Hilbert scan and the rest of the image is scanned in a Hilbert-like fashion. The modified Hilbert scan for the 8×8 blocks at level $n_0 = 3$ used for the QCIF video sequences is displayed in Fig. 5.5. Note that this scan is still optimal but for blocks larger than 16×16 the regions in the extended part of the modified Hilbert curve are not necessarily square.

It is well known that the DC values of the luminance (Y) and the chrominance channels (Cb, Cr) of neighboring blocks are highly correlated and therefore an encoding scheme should take advantage of this. One way of exploiting this fact is by encoding the DC values of consecutive intra blocks by first order DPCM. Therefore an additional dependency has been introduced and $r(x_{j-1}, x_j)$ is now equal to

$$r(x_{j-1}, x_j) = R^{SEG}(x_j) + R^{QDFD}(x_j) + R^{DVF}(x_{j-1}, x_j) + R^{DC}(x_{j-1}, x_j), \tag{7.15}$$

where $R^{DC}(x_{j-1}, x_j)$ is zero whenever the blocks associated with x_{j-1} and x_j are not both Intra coded, and equal to the number of bits needed to encode the difference between the DC coefficients of the two Intra coded blocks, otherwise. The DPCM of the DC values is extremely important for an Intra frame. Hence by using this DPCM the proposed coder will also efficiently encode an Intra frame, such as the first frame of a sequence.

From a theoretical point of view, every possible motion vector of block $b_{l,i}$ should be included in the set $M_{l,i}$. This means that for a typical search window of ± 15.5 pixels and an accuracy of 1/2 pixel, $|M_{l,i}| = 63 * 63 = 3969$, which is quite large. Most of these motion vectors, however, are not likely candidates for the optimal path, since they do not correspond well to the real motion in the scene and therefore they lead to a high distortion and a high rate.

To make the optimization process faster, such prior knowledge should be taken into account. Even though this is complicated in general, it can be easily achieved in the presented framework of DP by reducing the set $M_{l,i}$ of admissible motion vectors of block $b_{l,i}$. In the presented implementation, an initial motion vector search is first conducted for the 8×8 blocks at level $n_0 = 3$ by doing a block matching with integer accuracy. The K integer motion vectors which lead to the best prediction are kept, where $K = 10$ for the presented implementation. Then the set $M_{n_0,i}$ is defined as the set which contains the K top integer motion vectors plus their half pixel neighbors. Since the Skip and Intra encoding modes imply a zero motion vector the resulting set of all admissible state values of block $b_{n_0,i}$, $S_{n_0,i}$, is defined as follows,

$$S_{n_0,i} = \{n_0\} \times \{i\} \times \{10\} \times \Big((\{\text{Skip,Intra}\} \times \{\vec{0}\}) \cup (\{\text{Prediction,Inter}\} \times M_{n_0,i}) \Big). \tag{7.16}$$

After the set of the admissible state values has been defined for the bottom layer (n_0), the sets of the admissible state values for higher level blocks can be defined recursively in a Bottom-Up fashion, by calling the function "states(l, i)" below with the following parameters: states(N,0).

states(l, i)
  if ($l = n_0 + 1$)
    $S_{l,i} = \{l\} \times \{i\} \times \{10\} \times$
    $\Big((\{\text{Skip,Intra}\} \times \{\vec{0}\}) \cup (\{\text{Prediction,Inter}\} \times (\bigcap_{j=0}^{3} M_{l-1,4*i+j}))\Big)$;
  else
    states($l - 1, 4 * i + 0$);
    states($l - 1, 4 * i + 1$);
    states($l - 1, 4 * i + 2$);
    states($l - 1, 4 * i + 3$);
    $S_{l,i} = \{l\} \times \{i\} \times \{10\} \times$
    $\Big((\{\text{Skip,Intra}\} \times \{\vec{0}\}) \cup (\{\text{Prediction,Inter}\} \times (\bigcap_{j=0}^{3} M_{l-1,4*i+j}))\Big)$;

In other words, a block $b_{l,i}$ only includes a motion vector in $S_{l,i}$ if the motion vector has been selected by all of its child nodes. This leads to the fact that for small blocks, many motion vectors are considered, but the bigger the block, the fewer the included motion vectors. This reflects the well known fact that small block sizes lead to small DFD energy but not very consistent motion vector fields, whereas bigger blocks lead to consistent vector fields, but the DFD energy can be quite high [16].

Our experiments have shown that for $K = 10$ this restriction of the search space does not lead to a performance loss but increases the speed of the algorithm significantly.

7.6 EXPERIMENTAL RESULTS

In this section, the results of the proposed coder are compared to test model 4 (TMN4) [47] of the H.263 standard. Note that the optimal coder, like TMN4, writes a bit stream which is uniquely decodable by our decoder. Hence the listed bit rates are the effective number of bits used and not an estimate of the entropy.

TMN4 is selected as the comparison coder since it is a very low bit rate coder. At very low bit rates, the side information, such as segmentation, motion vectors and quantizers, takes up a bigger percentage of the overall bit rate than for a higher bit rate coder. Therefore the performance difference between the optimal coder and a coder which uses a fixed block size and encoding heuristics should be more pronounced. The optimal coder uses the same block-based

DCT encoding of the DFD as TMN4 and it has the same range for the motion estimation which leads to a fair comparison between the two coders.

In order to compare TMN4 and the optimal coder, TMN4 was used to encode every 4th of the first 200 frames of the QCIF sequence "Mother and Daughter" with a fixed quantizer step size $QP = 10$. The first frame was Intra coded with the above quantizer step size, for both TMN4 and the optimal coder. Since the "Mother and Daughter" sequence is considered to be recorded at 30 frames/second, this leads to an encoded frame rate of 7.5 frames/second. Since the optimal coder implementation encodes only the luminance component (Y) of the sequence, TMN4 was slightly changed so that it also encodes the Y channel only. The resulting rate and distortion were used for the comparison between TMN4 and the optimal coder. The employed distortion measure is the peak signal to noise ratio (PSNR) (see Eq. 6.28).

7.6.1 Matched distortion

The goal of this experiment is to compare the optimal coder with TMN4 in the case where their frame distortions are matched. This can be achieved by setting D_{max}, the maximum frame distortion from Eq. (7.9) equal to the frame distortion of TMN4. Clearly D_{max} changes from frame to frame, following the distortion profile of the TMN4 run. The optimal coder will lead to the smallest number of bits needed to encode a given frame for the given maximum distortion D_{max}.

The resulting rate and distortion per frame are displayed in Figs. 7.2, 7.3, 7.4 and 7.5. The average rate and distortion for the entire sequence are listed in Table 7.1 (second row). The distortion of the proposed coder follows the TMN4 distortion extremely closely. Clearly the proposed coder is superior to TMN4 when their frame distortions are matched, based on the resulting difference (5.73 kbits/s on the average) in bit rates.

Fig. 7.6 shows the reconstructed 12th frame of the sequence which is used to predict the 16th frame. The optimal QT segmentation, the optimal encoding mode selection and the optimal motion vector field are displayed in Figs. 7.7 and 7.8 for the 16th frame of the "Mother and Daughter" sequence. It is evident in Fig. 7.7 that smaller blocks are used at object boundaries where the motion vector field is discontinuous and larger blocks are used in background areas which are stationary. Note in Fig. 7.7 how the new object (hand) and the uncovered areas (left of the hand) are intra coded and the stationary back-

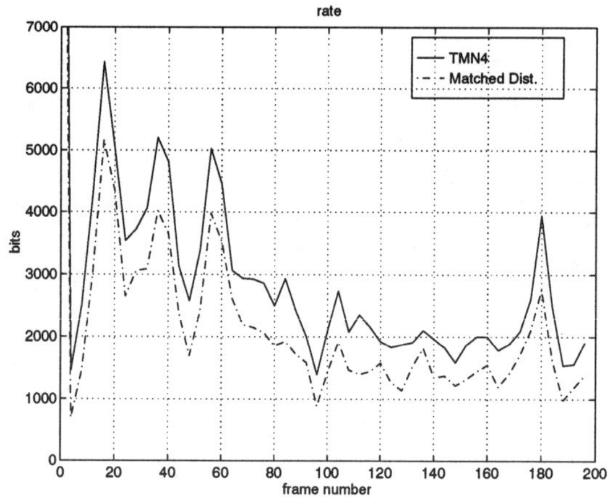

Figure 7.2 Rate comparison between TMN4 and the optimal coder, where the TMN4 distortion is the target distortion of the optimal coder.

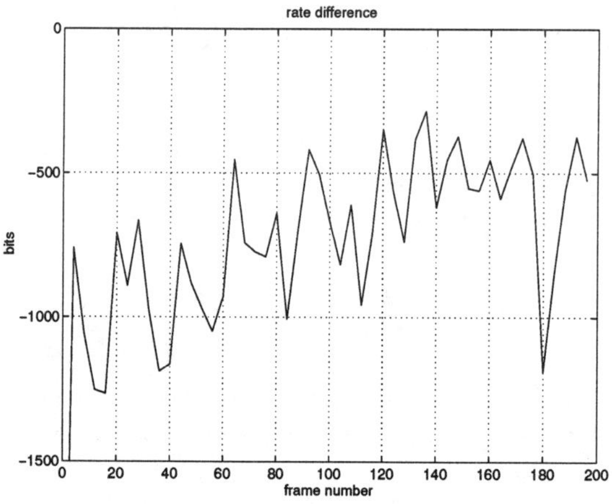

Figure 7.3 Rate difference between TMN4 and the optimal coder, where the TMN4 distortion is the target distortion of the optimal coder.

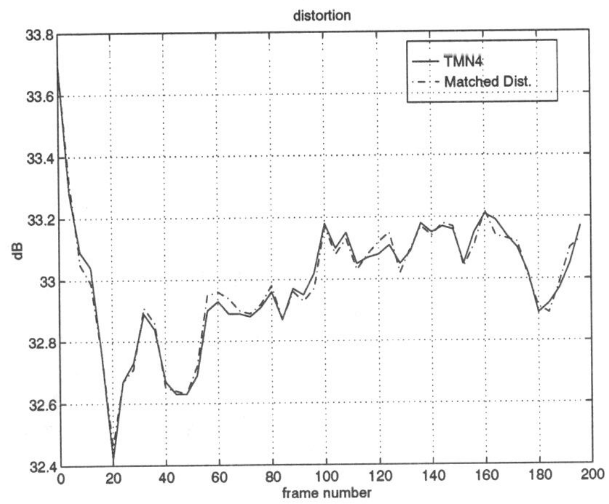

Figure 7.4 Distortion comparison between TMN4 and the optimal coder, where the TMN4 distortion is the target distortion of the optimal coder.

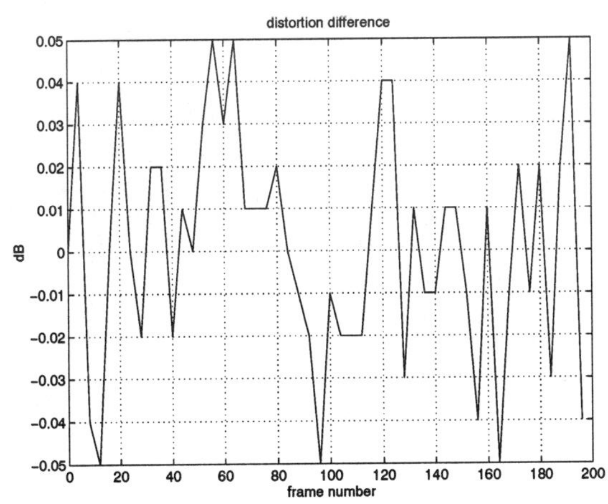

Figure 7.5 Distortion difference between TMN4 and the optimal coder, where the TMN4 distortion is the target distortion of the optimal coder.

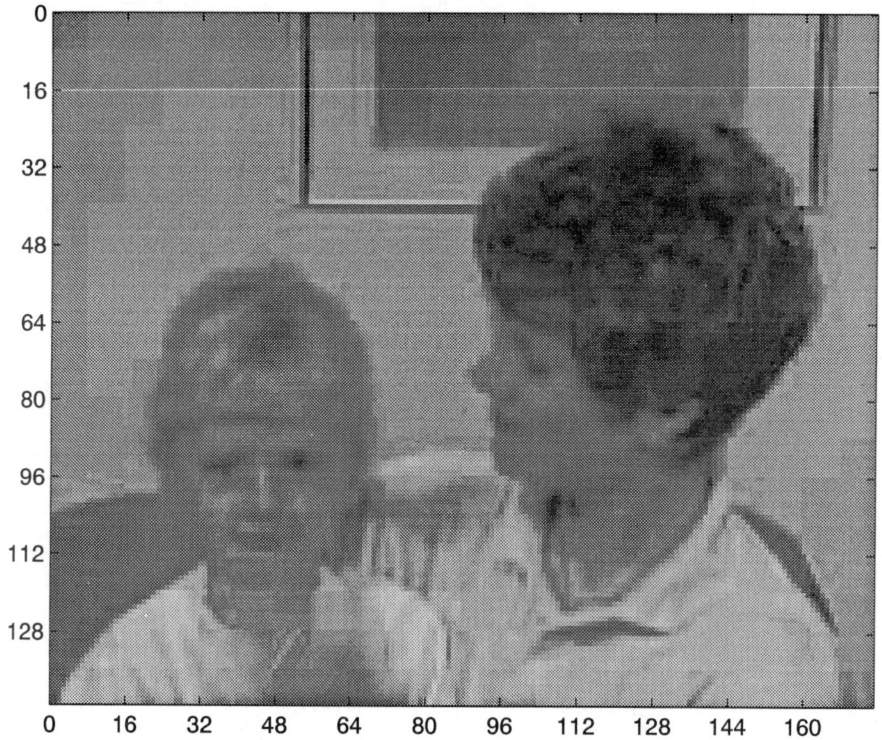

Figure 7.6 The 12th reconstructed frame of the "Mother and Daughter" sequence. This frame is used to predict the 16th frame.

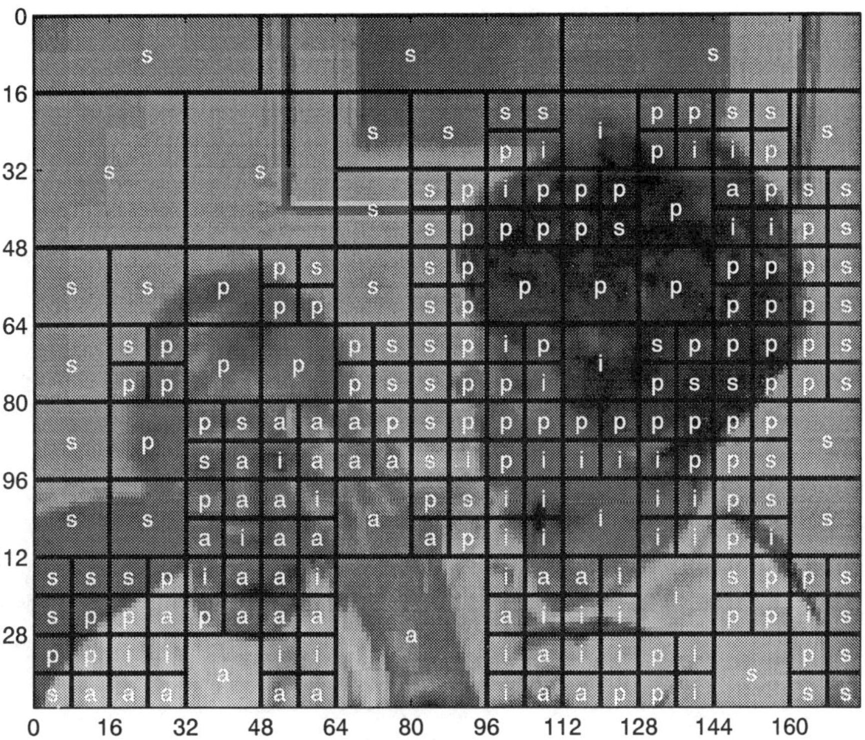

Figure 7.7 The optimal mode selection for the 16th frame of the "Mother and Daughter" sequence. (i) Inter mode, (s) Skip mode, (p) Prediction mode and (a) Intra mode.

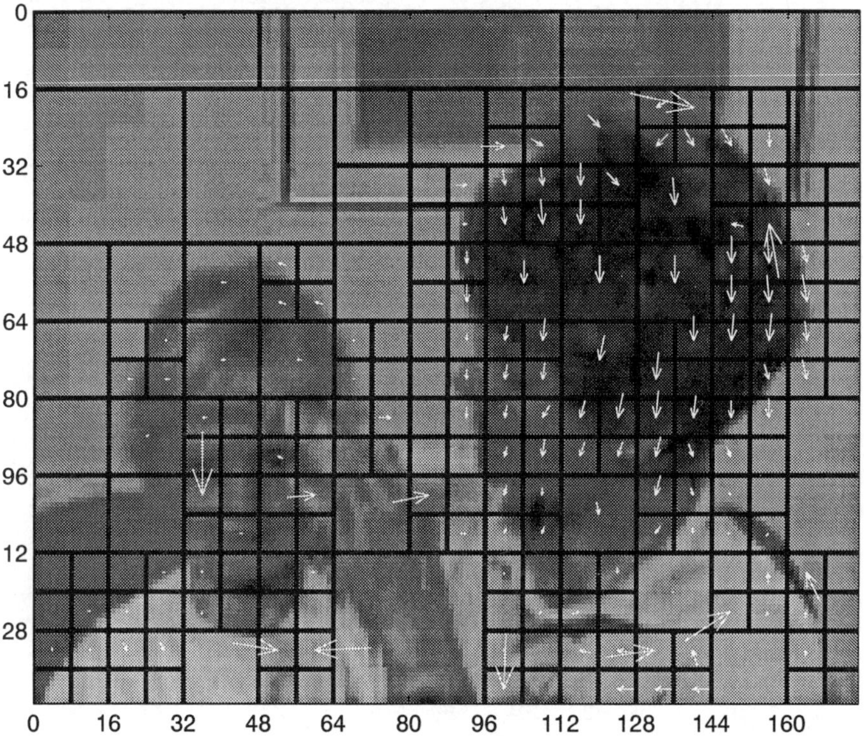

Figure 7.8 The optimal motion vector field for the 16th frame of the "Mother and Daughter" sequence.

	average rate kbits/s	average distortion (average PSNR) dB
TMN4	22.80	32.99
matched distortion	17.13	32.99
matched rate	22.30	33.71
constant distortion	14.52	32.43

Table 7.1 Average rate distortion comparison for the "Mother and Daughter" sequence between TMN4 and the proposed optimal coder for different modes of operation

ground is replaced by the blocks from frame 12 (Skip mode). As can be seen in Fig. 7.7, the Skip and the Prediction modes are the most commonly used encoding modes. In fact, for the entire sequence, without the first image, the Skip mode is used for 57.4%, the Prediction mode for 27.0%, the Inter mode for 14.1% and the Intra mode for 1.5% of the image. In other words, the DFD is only sent for 15.6% of the image, whereas for 84.4% of the image no DFD is sent. This is achieved by the superior motion compensation of this variable block size approach.

7.6.2 Matched rate

In this experiment, the proposed coder is compared to TMN4 in the case where their frame rates are matched. This can be achieved by setting R_{max}, the maximum frame rate from Eq. (7.5) equal to the frame rate of TMN4. Clearly R_{max} changes from frame to frame, following the rate profile of TMN4 and the proposed coder will minimize the resulting frame distortion for the given frame rate.

The resulting rate and distortion per frame are displayed in Figs. 7.9, 7.10, 7.11 and 7.12. The average rate and distortion for the entire sequence are shown in Table 7.1 (third row). Again, note that the rate of the proposed coder follows the TMN4 rate very closely. Besides being able to outperform TMN4 for matched rates, this experiment also shows the enormous potential of this approach with respect to rate control since the optimal coder can follow an

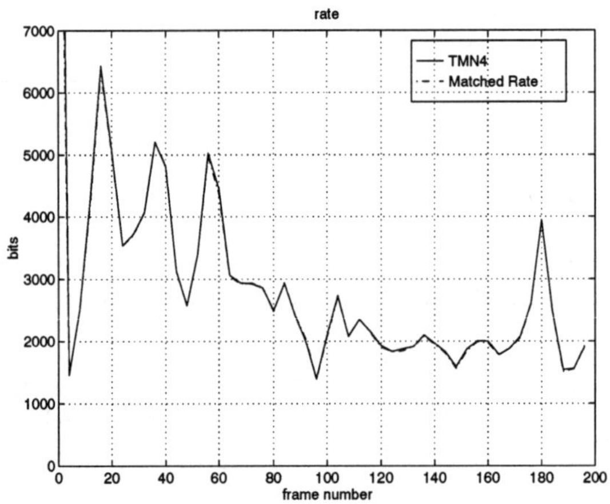

Figure 7.9 Rate comparison between TMN4 and the optimal coder, where the TMN4 rate is the target rate of the optimal coder.

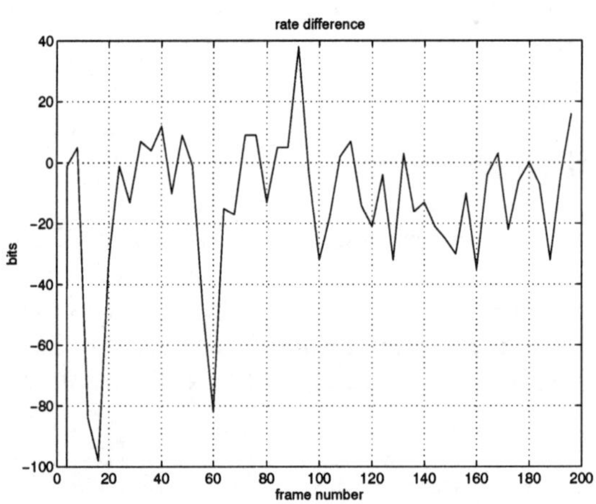

Figure 7.10 Rate difference between TMN4 and the optimal coder, where the TMN4 rate is the target rate of the optimal coder.

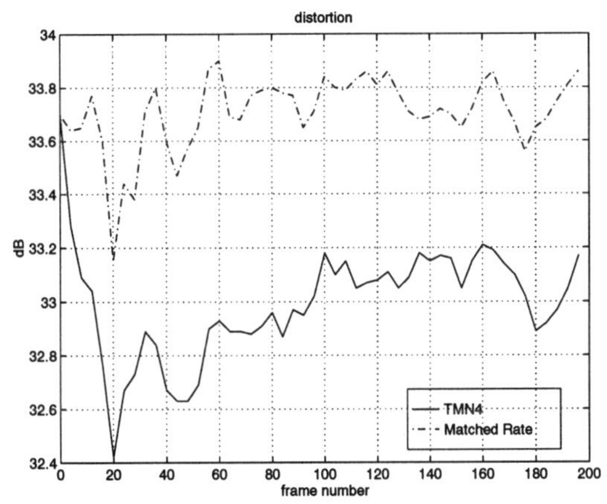

Figure 7.11 Distortion comparison between TMN4 and the optimal coder, where the TMN4 rate is the target rate of the optimal coder.

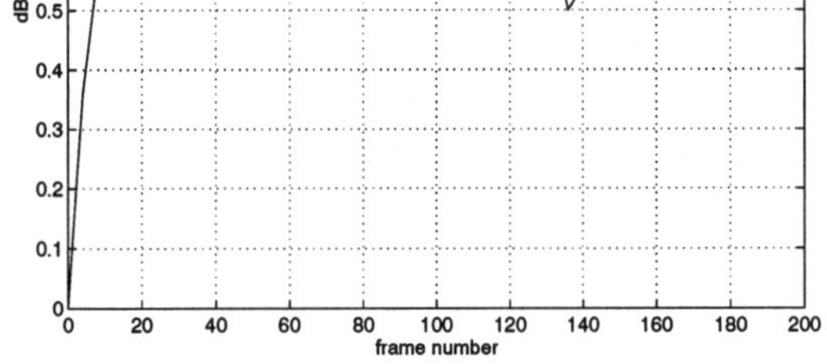

Figure 7.12 Distortion difference between TMN4 and the optimal coder, where the TMN4 rate is the target rate of the optimal coder.

arbitrary bit assignment per frame and produce the smallest possible distortion for the given bit budget.

7.6.3 Constant distortion

So far TMN4 and the proposed coder have been compared in terms dictated by the TMN4 run. This gives TMN4 an advantage, since it can set the rate distortion profile which then has to be followed by the proposed coder.

An interesting application of the proposed coder is for channels which can accept a variable bit rate, such as an ATM network. For such applications, one would like to keep the distortion constant which can be achieved by setting D_{max}, the maximum frame distortion from Eq. (7.9), equal to the desired frame distortion. The proposed coder allows therefore for the encoding of a video sequence with constant frame quality.

For the experiment discussed next, D_{max} was selected to be equal to the minimum frame PSNR of the TMN4 run, since this is the best quality TMN4 can guarantee over the entire sequence. In this experiment, the distortion is constant for each frame of the sequence, and hence the average distortion is equal to the frame distortion, which is equal to the maximum frame distortion of TMN4. Therefore the fixed quantizer step size QP, which is equal to 10 in the TMN4 run, is set to 11 for the constant distortion experiment, since a coarser quantizer step size is required for a higher average distortion.

The resulting rate and distortion per frame are shown in Figs. 7.13, 7.14, 7.15 and 7.16. The average rate and distortion for the entire sequence are shown in Table 7.1 (fourth row). Clearly the goal of constant distortion (quality) has been achieved and the resulting average rate is much lower than the TMN4 rate, even though visually these two encoded sequences cannot be distinguished. Some observers even prefer the constant quality sequence over the TMN4 sequence. One possible explanation for this fact is based on the globally optimal selection of the DFD and the DVF. Recall that a Hilbert scan is used for the DPCM encoding of the DVF and a smooth DVF along this path leads to a low bit rate. Hence the optimal solution enforces a global smoothness constraint on the DVF which in turn leads to predicted frames which are more visually pleasing than the ones block matching produces.

So far all the experiments presented have been based on the "Mother and Daughter" sequence. In the field of very low bit rate video coding, a common

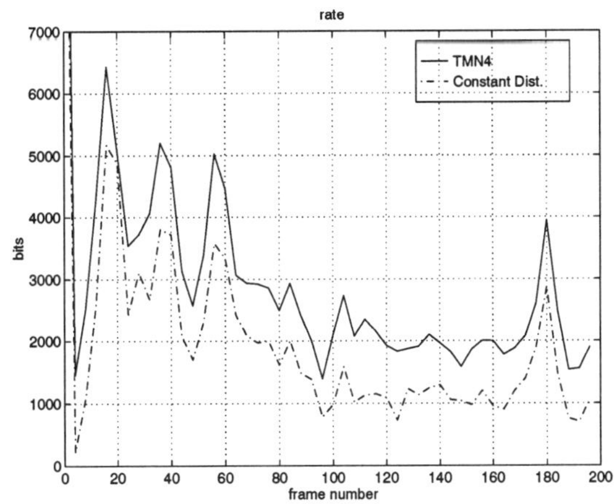

Figure 7.13 Rate comparison between TMN4 and the optimal coder, where the distortion of the optimal coder is fixed.

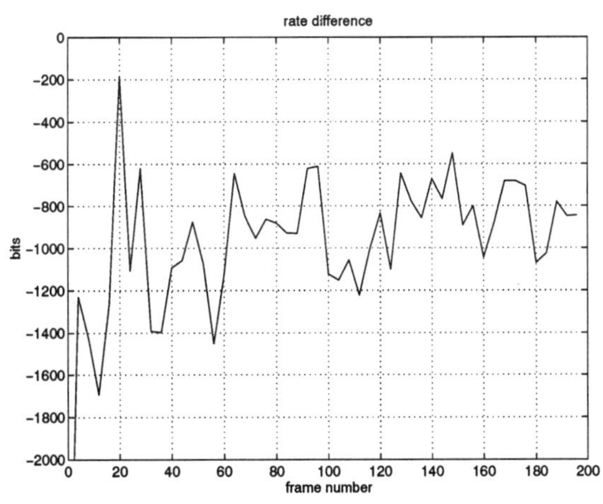

Figure 7.14 Rate difference between TMN4 and the optimal coder, where the distortion of the optimal coder is fixed.

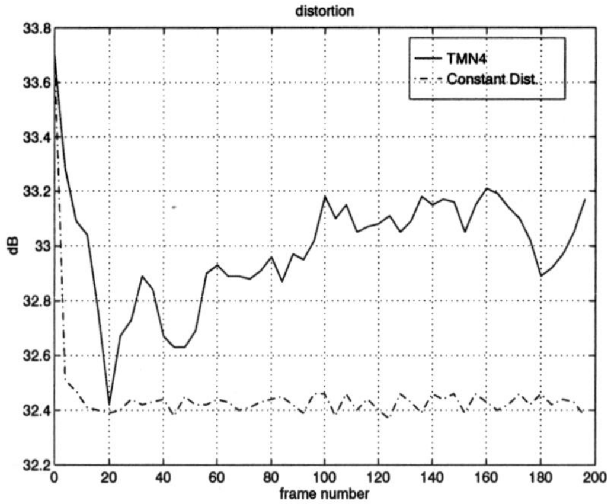

Figure 7.15 Distortion comparison between TMN4 and the optimal coder, where the distortion of the optimal coder is fixed.

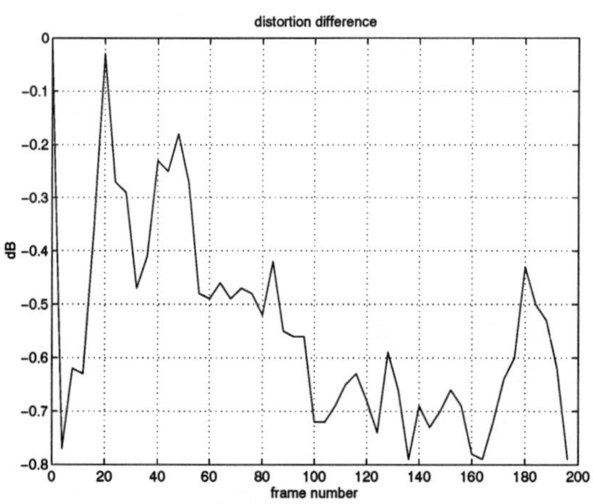

Figure 7.16 Distortion difference between TMN4 and the optimal coder, where the distortion of the optimal coder is fixed.

test sequence is the "Miss America" sequence. It is well known that the motion in this sequence is almost completely translational and therefore this sequence should be well modeled by the proposed variable block size encoder. Again, the "Miss America" sequence is a QCIF sequence of which every 4th luminance (Y) frame is encoded which leads to an encoded frame rate of 7.5 frames per second. The sequence is coded with a quantizer step size of 12 and the target distortion is 35.0 dB PSNR. The resulting average bit rate is 8.30 kbits per second and the average distortion is 35.0 dB PSNR. The above results are for our QCIF "Miss America" sequence. Unfortunately, the original "Miss America" sequence is in the common intermediate format (CIF) (352×288 pixels) and there are different ways of converting it to a QCIF sequence. Consequently there exist different QCIF "Miss America" sequences, with different high frequency contents. We obtained a popular version of a QCIF "Miss America" sequence from Norwegian Telecom (TELENOR) and it turned out that this sequence is not as sharp as the one we had available. This second "Miss America" sequence was compressed at 35.0 dB PSNR using only 5.70 kbits per second.

This extremely low bit rate for this high PSNR value can be explained by looking at the encoding modes. The prediction of the future frames is so successful that the DFD has to be sent for less than 6% of the sequence. In other words, the Intra or Inter modes are only used to update the scene for less than 6% of the total image size. These 6% can be considered model failure regions and then they can be directly compared to the 4% model failure regions in the object-oriented approach presented in [73]. As expected, the Intra and Inter modes are most often used around the eyes and the mouth.

Fig. 7.17 shows the optimal quad-tree segmentation and the encoding modes of frame 80 of the sharper "Miss America" sequence and Fig. 7.18 shows the optimal inhomogeneous motion vector field.

We also compared our algorithm to the most recent test model for the H.263 standard, TMN6 [46]. The differences between TMN4 and TMN6 are described in section 6.8.3. Table 7.2 shows, the proposed scheme still outperforms H.263, even in its most advanced form.

7.7 SUMMARY

In this chapter, we have presented a general theory for the optimal bit allocation among QT segmentation, displacement vector field (DVF) and displaced

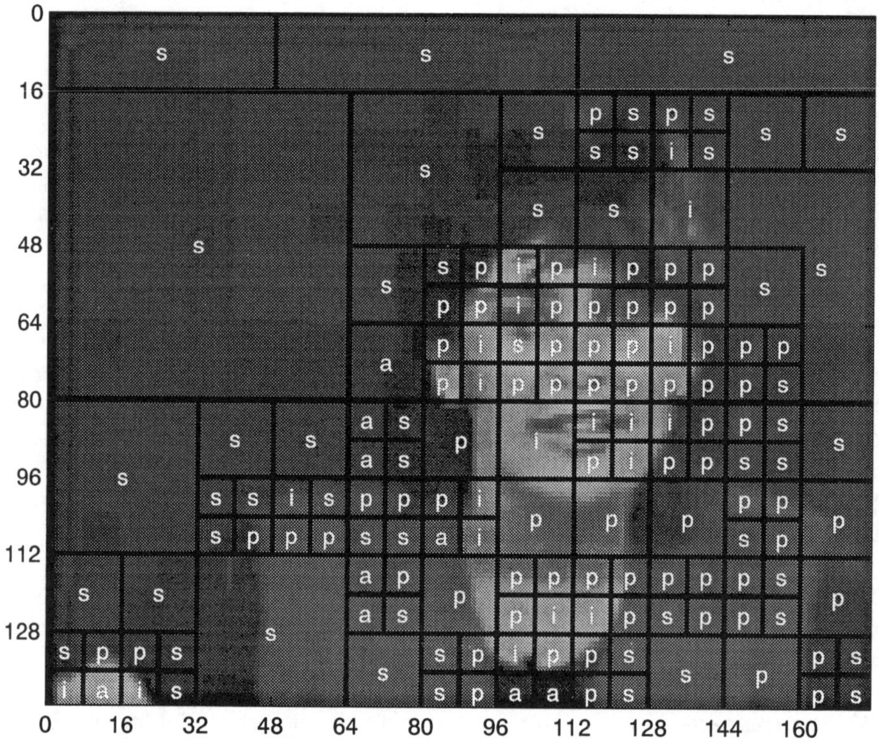

Figure 7.17 Optimal quad-tree segmentation and encoding modes for the 80th frame of the "Miss America" sequence

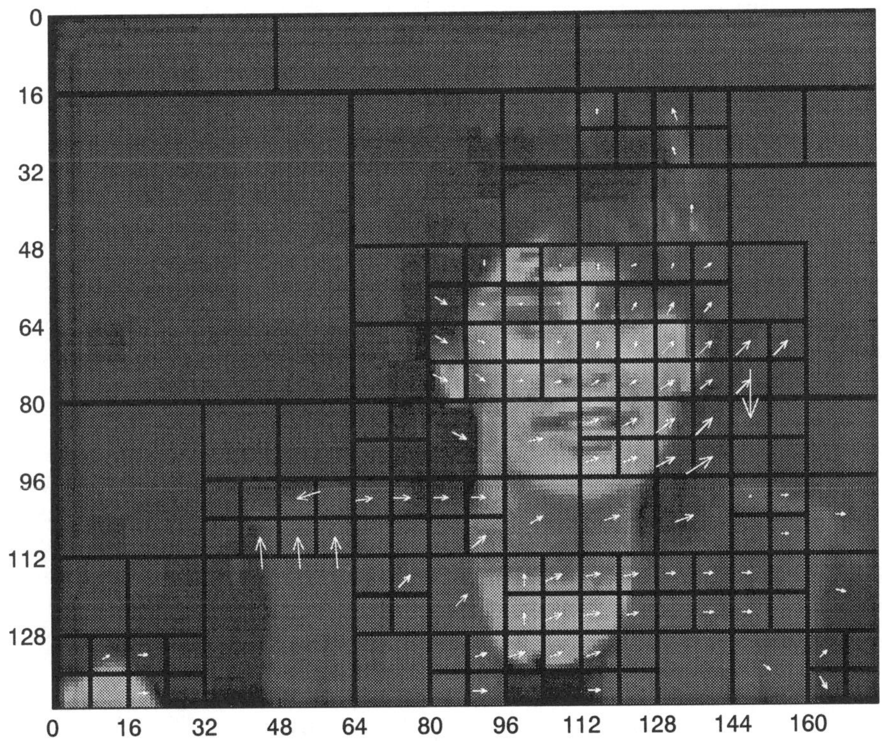

Figure 7.18 Optimal inhomogeneous motion vector field for the 80th frame of the "Miss America" sequence

	average rate kbits/s	average distortion (average PSNR) dB
TMN6	20.2	32.9
matched distortion	17.6	32.9
matched rate	20.2	33.5
constant distortion	15.5	32.3

Table 7.2 Average rate distortion comparison for the "Mother and Daughter" sequence between TMN6 and the proposed optimal coder for different modes of operation

frame difference (DFD). The theory can be applied to any variable block size motion compensated video coder (VBSMCVC), where the variable block sizes are encoded using a QT structure, the DVF is encoded by first order DPCM, the DFD is encoded by a block-based scheme and an additive distortion measure is employed.

We first considered a lossless VBSMCVC and derived the optimal bit allocation algorithm which is based on dynamic programming (DP). We then addressed the problem of lossy VBSMCVC and we showed that Lagrangian relaxation and DP can be used to find the convex hull approximation to the optimal solution for the minimum total distortion criterion. We also showed how the lossless solution can be used to solve the optimal bit allocation problem for the minimum maximum distortion criterion.

A DCT-based implementation of an optimal coder has then been discussed and the results of this coder have been presented. We showed that this DCT-QT coder can outperform TMN4 and TMN6 in the rate-distortion sense.

Besides being able to outperform TMN4 and TMN6, the proposed coder is well suited for a rate control scheme, since for any given bit rate, it will encode the current frame with the smallest possible distortion. Also, the optimal solution enforces a global smoothness constraint on the DVF which in turn leads to predicted frames which are more visually pleasing than the ones produced by block matching. Hence the encoded sequences are not only superior in a rate distortion sense but also in a subjective sense.

8

AN OPTIMAL POLYGONAL BOUNDARY ENCODING SCHEME

In this chapter, we present fast and efficient methods for the lossy encoding of object boundaries which are given as 8-connect chain codes [180, 182, 177, 179]. We approximate the boundary by a polygon and consider the problem of finding the polygon which leads to the smallest distortion for a given number of bits. We also address the dual problem of finding the polygon which leads to the smallest bit rate for a given distortion. We consider two different classes of distortion measures. The first class is based on the maximum operator and the second class is based on the summation operator. For the first class, we derive a fast and optimal scheme which is based on a shortest path algorithm for a weighted directed acyclic graph. For the second class we propose a solution approach which is based on the Lagrangian multiplier method, which uses the above mentioned shortest path algorithm. Since the Lagrangian multiplier method can only find solutions on the convex hull of the operational rate distortion function, we also propose a tree pruning algorithm which can find all optimal solutions. Finally we present results of the proposed schemes using objects from the "Miss America" sequence.

8.1 INTRODUCTION

The encoding of planar curves is an important problem in many different fields, such as CAD, object recognition and object-oriented video coding. This research is motivated by object-oriented video coding, but the developed algorithms can also be used for other applications. A major problem in object-oriented video coding [126, 73] is the efficient encoding of object boundaries. There are two main approaches for encoding the segmentation information:

a lossy approach, which is based on a spline approximation of the boundary [86, 73], and a lossless approach which is based on chain codes [55]. The proposed boundary encoding scheme is a lossy scheme which can be considered a combination of the spline and the chain code approaches, since the boundary is approximated by a polygon and its vertices are encoded relative to each other. The approximation of the boundary by a polygon is similar to the spline approximation approach, whereas the encoding of the successive vertices is achieved with a chain code-like scheme.

In [86] B-spline curves are used to approximate a boundary. An optimization procedure is formulated for finding the optimal locations of the control points by minimizing the mean squared error between the boundary and the approximation. This is an appropriate objective when the smoothing of the boundary is the main problem. When, however, the resulting control points need to be encoded, the tradeoff between the encoding cost and the resulting distortion needs to be considered. By selecting the mean squared error as the distortion measure and allowing for the location of the control points to be anywhere on the plane, the resulting optimization problem is continuous and convex and can be solved easily. In order to encode the positions of the resulting control points efficiently, however, one needs to quantize them, and therefore the optimality of the solution is lost. It is well known that the optimal solution to a discrete optimization problem (quantized locations) does not have to be close to the solution of the corresponding continuous problem.

In [73] a boundary is approximated by a combination of splines and polygons, where the distortion measure employed is the maximum distance between the approximation and the original boundary. First the vertices of the polygon are heuristically found and then these vertices are used to find a spline representation of the boundary. For every segment where the spline representation does not exceed the maximum distance from the original boundary, the spline representation is used instead of the polygon. This leads to a smoother approximation. Again, the approach is well suited for the smoothing of the original boundary and there is also an inherent control over the maximum distance between the approximation and the original boundary, but the resulting rate is not explicitly taken into account. If each control point is encoded using a fixed length codeword, then minimizing the number of control points is equivalent to minimizing the resulting bit rate. On the other hand, if the control points are along a natural boundary, there exists a high correlation between the location of two consecutive points, and a predictive encoding scheme should be employed instead of a fixed length codeword scheme. In any case this approach does not facilitate the rate constrained encoding of a boundary, i.e., the rate is given and one wants to find the best approximation.

Freeman [55] originally proposed the use of chain coding for boundary quantization and encoding, which has attracted considerable attention over the last thirty years [173, 93, 134, 89, 157]. The most common chain code is the 8-connect chain code which is based on a rectangular grid superimposed on a planar curve. The curve is quantized using the grid intersection scheme [55] and the quantized curve is represented using a string of increments. Since the planar curve is assumed to be continuous, the increments between grid points are limited to the 8 grid neighbors, and hence an increment can be represented by 3 bits. There have been many extensions to this basic scheme such as the generalized chain codes [173], where the coding efficiency has been improved by using links of different length and different angular resolution. In [89] a scheme is presented which utilizes patterns in a chain code string to increase the coding efficiency and in [157] differential chain codes are presented, which employ the statistical dependency between successive links. There has also been interest in the theoretical performance of chain codes. In [93] the performance of different quantization schemes is compared, whereas in [134] the rate distortion characteristics of certain chain codes are studied. In this paper, we are not concerned with the quantization of the continuous curve, since we assume that the object boundaries are given with pixel accuracy.

Most boundaries contain many straight lines or lines with a very small curvature, which result in runs of the same increment. Therefore a run-length encoding scheme can be used to encode the 8-connect chain code even more efficiently. Clearly the larger the number of straight lines a boundary contains, the more efficient a chain code/run-length scheme is. This is the idea behind some preprocessing algorithms [146, 206] which are used to "straighten" the boundary, i.e., these algorithms lead to a new boundary which can be encoded more efficiently. Clearly such a scheme is a lossy chain code/run-length encoding scheme since this preprocessing introduces an error in the boundary representation. The introduction of an error is usually permissible, as long as the visual distortion is considered insignificant. The main problem with these approaches is that there is at best an indirect control over the resulting rate distortion tradeoff.

In this chapter, we present algorithms where the preprocessing step and the 8-connect chain code/run-length encoding are combined into an optimal lossy segmentation encoding scheme. The proposed approach offers complete control over the tradeoff between distortion and bit rate. Note that this is achieved in an optimal fashion, resulting in an efficient encoding scheme.

This chapter is organized as follows: In section 8.2 we define the problem and introduce the required notation and assumptions. In section 8.3 we focus on a first class of distortion measures which are based on the maximum operator, such as the maximum absolute distance. First we consider the problem of finding the polygon which requires the smallest bit rate for a given distortion. We solve this problem by introducing a scheme which is based on a shortest path algorithm for a weighted directed acyclic graph. We then consider the dual problem, of finding the polygon with the smallest distortion for a given bit rate. We derive an iterative scheme which employs the shortest path algorithm and prove that this scheme converges to the global optimum. In section 8.4 we focus on a second class of distortion measures which are based on the summation operator, such as the mean squared distance. Again, we consider the problem of finding the polygon which requires the smallest bit rate for a given distortion. We show how the Lagrangian multiplier method can be employed to transform this constrained problem into a series of unconstrained problems which can be solved using the shortest path algorithm introduced earlier. Like every Lagrangian-based approach, the proposed scheme results only in solutions which belong to the convex hull of the operational rate distortion function. Hence we also propose a second algorithm, which is based on a tree pruning scheme, which can find all optimal solutions. Since both of these algorithms are symmetric in the rate and the distortion, the dual problem, that of finding the polygon which results in the smallest distortion for a given bit rate, can also be solved by the same algorithms. In section 8.5 we propose modifications to the above schemes so that when there is more than one optimal polygon, a secondary distortion measure can be used to select the best polygon from the set of optimal polygons. In section 8.6 an extension is introduced which results in more freedom in selecting the vertices of the optimal polygon. In section 8.7 we extend the results for the jointly optimal encoding of multiple boundaries. In section 8.8 we introduce different vertex encoding schemes which are based on an 8-connect chain code and run-length coding. In section 8.9 we present results of the proposed algorithms and in section 8.10 we summarize this chapter.

8.2 PROBLEM FORMULATION

In this section we introduce the required notation, state the underlying assumptions and formulate the problem mathematically. The main idea behind the proposed approach is to approximate a given boundary by a polygon, and to encode the polygons vertices instead of the original boundary. Since we assume that the original boundary is represented with pixel accuracy, it can be losslessly encoded by an 8-connect chain-code. We propose to approximate the boundary with a low order polygon which can be encoded efficiently.

The following notation will be used. Let $B = \{b_0, \ldots, b_{N_B-1}\}$ denote the connected boundary which is an ordered set, where b_j is the j-th point of B and N_B is the total number of points in B. Note that in the case of a closed boundary, $b_0 = b_{N_B-1}$. Let $P = \{p_0, \ldots, p_{N_P-1}\}$ denote the polygon used to approximate B, which is also an ordered set, with p_k the k-th vertex of P, N_P the total number of vertices in P and the k-th edge starts at p_{k-1} and ends at p_k. Since P is an ordered set, the ordering rule and the set of vertices uniquely define the polygon. Later, we will elaborate on the fact that the polygon is an ordered set.

We assume that the vertices of the polygon are encoded differentially which is an efficient method for natural boundaries since the location of the current vertex is strongly correlated with the location of the previous one. We denote the required bit rate for the differential encoding of vertex p_k given vertex p_{k-1} by $r(p_{k-1}, p_k)$. Hence the bit rate $R(p_0, \ldots, p_{N_P-1})$ for the entire polygon is,

$$R(p_0, \ldots, p_{N_P-1}) = \sum_{k=0}^{N_P-1} r(p_{k-1}, p_k), \tag{8.1}$$

where $r(p_{-1}, p_0)$ is set equal to the number of bits needed to encode the absolute position of the first vertex. For a closed boundary, i.e., the first vertex is identical to the last one, the rate $r(p_{N_P-2}, p_{N_P-1})$ is set to zero since the last vertex does not need to be encoded. Note that this rate depends on the specific vertex encoding scheme. We present two such schemes which are a combination of an 8-chain code and a run-length scheme in section 8.8.

In general the polygon which is used to approximate the boundary could be permitted to place its vertices anywhere on the plane. We restrict the vertices to belong to the original boundary ($p_k \in B$), so that we can employ a fast polygon selection algorithm. This restriction results in the following fact, which we employ to derive low complexity optimization algorithms. The k-th polygon edge which connects two consecutive vertices, p_{k-1} and p_k, is an approximation to the partial boundary $\{b_j = p_{k-1}, b_{j+1}, \ldots, b_{j+l} = p_k\}$, which contains $l+1$ boundary points. Therefore, we can measure the quality of this approximation by an edge distortion measure which we denote by $d(p_{k-1}, p_k)$. The polygon distortion measure can then be expressed as the sum or the maximum of all edge distortion measures. In section 8.6, we present a scheme which relaxes this restriction by permitting the vertices to belong to a "band" around the original boundary. The resulting scheme is very efficient but computationally more expensive.

There are several different distortion measures which can be employed. One popular distortion measure for curve approximations is the maximum absolute distance, which has also been employed in [93, 134, 73, 180]. Besides its perceptual relevance, this distortion measure has the advantage that it can be computed efficiently. Let $d(p_{k-1}, p_k, t)$ be the shortest absolute distance between the line which goes through p_{k-1} and p_k and an arbitrary point t. This distance can be expressed as follows,

$$d(p_{k-1}, p_k, t) = \frac{|(t_x - p_{k-1,x})(p_{k,y} - p_{k-1,y}) - (t_y - p_{k-1,y})(p_{k,x} - p_{k-1,x})|}{\sqrt{(p_{k,x} - p_{k-1,x})^2 + (p_{k,y} - p_{k-1,y})^2}} \tag{8.2}$$

where the subscripts x and y indicate the x and y coordinates of a particular point. Then the maximum absolute distance between the partial boundary $\{b_j = p_{k-1}, b_{j+1}, \ldots, b_{j+l} = p_k\}$ and the edge (p_{k-1}, p_k) is given by,

$$d(p_{k-1}, p_k) = \max_{t \in \{b_j = p_{k-1}, b_{j+1}, \ldots, b_{j+l} = p_k\}} d(p_{k-1}, p_k, t). \tag{8.3}$$

Another popular distortion measure is the mean squared distance (error), which has been used in [86, 182] and is of the following form,

$$d(p_{k-1}, p_k) = \frac{1}{N_B} \sum_{t \in \{b_j = p_{k-1}, b_{j+1}, \ldots, b_{j+l} = p_k\}} d^2(p_{k-1}, p_k, t). \tag{8.4}$$

So far we have only discussed the edge distortion measures, i.e., the measures which judge the approximation of a certain partial boundary by a given polygon edge. In general we are interested in a polygon distortion measure which can be used to determine the quality of approximation of an entire polygon. We will treat two different classes of polygon distortion measures. The first class is based on the maximum operator (or equivalently, on the minimum operator) and is of the following form,

$$D(p_0, \ldots, p_{N_P-1}) = \max_{k \in [0, \ldots, N_P-1]} d(p_{k-1}, p_k), \tag{8.5}$$

where $d(p_{-1}, p_0)$ is defined to be zero. We will denote all distortion measures based on the above definition as class one distortion measures.

The second class of distortion measures, which is based on the summation operator, is of the following form,

$$D(p_0, \ldots, p_{N_P-1}) = \sum_{k=0}^{N_P-1} d(p_{k-1}, p_k), \tag{8.6}$$

where again $d(p_{-1}, p_0)$ is set equal to zero. We will denote all distortion measures based on the above definition as class two distortion measures.

If we select the maximum absolute distance as the polygon distortion measure, then we have to use Eq. (8.3) for the edge distortion and Eq. (8.5) for the polygon distortion. On the other hand, if we select the mean squared distance as the distortion measure, then we have to use Eq. (8.4) for the edge distortion and Eq. (8.6) for the polygon distortion. Note that there are many other polygon distortion measures which fit into this framework, such as the absolute area or the total number of pixels between the boundary and the polygon.

As we mentioned above, one of the big advantages of restricting the vertices of the polygon to belong to the original boundary is the ability to express the polygon distortion as the sum or the maximum of the edge distortions. We will employ this fact later to derive fast polygon selection algorithms. These fast algorithms are necessary since the number of possible polygons is extremely high. If we define the smallest possible polygon as a single point, then, given the degree of the polygon (N_P), there are $\frac{N_B!}{(N_B-N_P)! * N_P!}$ different selections of N_P vertices from the original boundary. Since we have defined the polygon to be an ordered set, the set of vertices uniquely specifies a polygon. The degree of the polygon (N_P) is also a variable, therefore the total number of possible polygons is equal to

$$\sum_{k=1}^{N_B} \frac{N_B!}{(N_B - k)! * k!}. \tag{8.7}$$

Clearly an exhaustive search is not a reasonable approach.

In the reminder of the paper we introduce fast and efficient algorithms for both classes of polygon distortion measures which solve the following constrained optimization problem,

$$\min_{p_0, \ldots, p_{N_P-1}} D(p_0, \ldots, p_{N_P-1}), \quad \text{subject to:} \quad R(p_0, \ldots, p_{N_P-1}) \leq R_{max}, \tag{8.8}$$

where R_{max} is the maximum bit rate permitted for the encoding of the boundary. We also present algorithms which solve the dual problem,

$$\min_{p_0,\ldots,p_{N_P-1}} R(p_0,\ldots,p_{N_P-1}), \quad \text{subject to:} \quad D(p_0,\ldots,p_{N_P-1}) \leq D_{max}, \tag{8.9}$$

where D_{max} is the maximum distortion permitted. Note that there is an inherent tradeoff between the rate and the distortion in the sense that a small distortion requires a high rate, whereas a small rate results in a high distortion. As we will see, the solution approaches for problems (8.8) and (8.9) are related in the sense that the algorithms are symmetric with respect to the rate and the distortion for distortion measures based on the summation operator. When the distortion measure is based on the maximum operator, then the algorithm developed to solve problem (8.9) is used iteratively to solve problem (8.8).

8.3 DISTORTION MEASURES BASED ON THE MAXIMUM OPERATOR

In this section we introduce two algorithms to solve the problems stated in Eqs. (8.8) and (8.9) for class one distortion measures, such as the maximum absolute distance.

8.3.1 The minimum rate case

First we consider the minimum rate case which is stated in Eq. (8.9). The goal of the proposed algorithm is to find the polygon whose vertices can be encoded with the smallest number of bits. This selection is constrained by the fact that the selected polygon must result in a distortion smaller than or equal to the maximum distortion.

The key observation for deriving an efficient search is the fact that given a certain vertex of a polygon (p_{k-1}) and the rate which is required to code the polygon up to and including this vertex ($R_{k-1}(p_{k-1})$), the selection of the next vertex p_k is independent of the selection of the previous vertices $p_0,\ldots,p_{k-2}$. This is true since the rate can be expressed recursively as a function of the segment rates $r(p_{k-1},p_k)$ and the segment distortions $d(p_{k-1},p_k)$. That is,

$$R_k(p_k) = R_{k-1}(p_{k-1}) + w(p_{k-1},p_k), \tag{8.10}$$

where

$$w(p_{k-1}, p_k) = \begin{cases} \infty & : \quad d(p_{k-1}, p_k) > D_{max} \\ r(p_{k-1}, p_k) & : \quad d(p_{k-1}, p_k) \leq D_{max} \end{cases} . \tag{8.11}$$

This recursion needs to be initialized by setting $R_{-1}(p_{-1})$ equal to zero. Note that by the definition of $w(p_{k-1}, p_k)$, the rate for a polygon which does not satisfy the maximum distortion constraint is infinite. Clearly $R_{N_P-1}(p_{N_P-1})$ $= R(p_0, \ldots, p_{N_P-1})$, the rate for the entire polygon.

As indicated above, we need to start the search for an optimal polygon at a given vertex. If the boundary is not closed, the first boundary point b_0 has to be selected as the first vertex p_0. For a closed boundary, we can select any boundary point. A good choice is the point with the highest curvature since it is the most likely point to be included in any polygonal approximation. For future convenience, we relabel the boundary so that the first vertex of the polygon p_0 coincides with the first point of the boundary b_0. Note that our experiments have shown that the efficiency of the solution approach does not depend on the choice of the initial starting point. Besides fixing the first vertex of the polygon, we also require that the last vertex p_{N_P-1} is equal to the last point of the boundary b_{N_B-1}. This leads to a closed polygonal approximation for a closed boundary. For a boundary which is not closed, this condition, together with the starting condition, makes sure that the approximation starts and ends at the same points as the boundary.

Using Eq. (8.10) the problem stated in Eq. (8.9) can be formulated as a shortest path problem in a weighted directed graph $G = (V, E)$, where V is the sets of graph vertices and E is the set of edges (see Fig. 8.1). Let $V = B$ since every boundary point can be a polygon vertex. Note that there are two kind of vertices, polygon vertices and graph vertices. In the proposed formulation, each graph vertex represents a possible polygon vertex and henceforth we will drop the distinction between these two entities. The edges between the vertices represent the line segments of the polygon. A directed edge is denoted by the ordered pair $(u, v) \in E$ which implies that the edge starts at vertex u and ends at vertex v. Since every combination of different boundary points can represent a line segment of a valid polygon, the edge set E is defined as follows, $E = \{(b_i, b_j) \in B^2 : \forall\ i \neq j\}$ (see Fig. 8.1). A path of order K from vertex u to a vertex u' is an ordered set $\{v_0, \ldots, v_K\}$ such that $u = v_0$, $u' = v_K$ and $(v_{k-1}, v_k) \in E$ for $k = 1, \ldots, K$. The order of the path is the number of edges in the path. The length of a path is defined as follows,

$$\sum_{k=1}^{K} w(v_{k-1}, v_k), \tag{8.12}$$

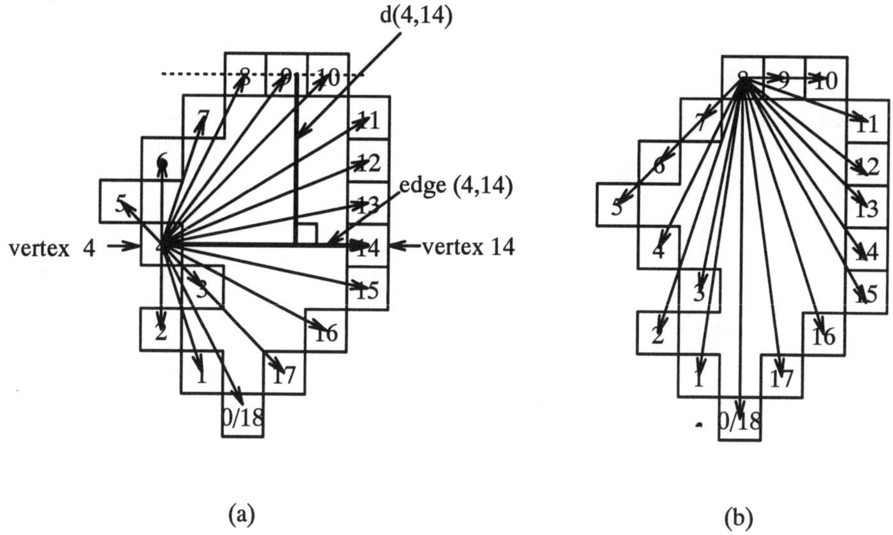

Figure 8.1 Interpretation of the boundary and the polygon approximation as a fully connected weighted directed graph. Note that the set of all edges E equals $\{(b_i, b_j) \in B^2 : i \neq j\}$. Two representative subsets are displayed: (a) $\{(b_4, b_j) \in B^2 : \forall j \neq 4\}$ and (b) $\{(b_8, b_j) \in B^2 : \forall j \neq 8\}$.

where $w(u,v) : E \rightarrow \mathcal{R}$ is a weight function defined in accordance with Eq. (8.11) as follows,

$$w(u,v) = \begin{cases} \infty & : \ d(u,v) > D_{max} \\ r(u,v) & : \ d(u,v) \leq D_{max} \end{cases} . \tag{8.13}$$

Again, note that the above definition of the weight function leads to a length of infinity for every path (polygon) which includes a line segment resulting in an approximation error larger than D_{max}. Therefore a shortest path algorithm will not select these paths. Every path which starts at vertex p_0 and ends at vertex p_{N_B-1} and does not result in a path length of infinity, results in a path length equal to the rate of the polygon it represents. Therefore the shortest of all those paths corresponds to the polygon with the smallest bit rate which is the solution to the problem in Eq. (8.9).

The classical algorithm for solving such a single-source shortest-path problem, where all the weights are non-negative, is Dijkstra's algorithm [34] with time

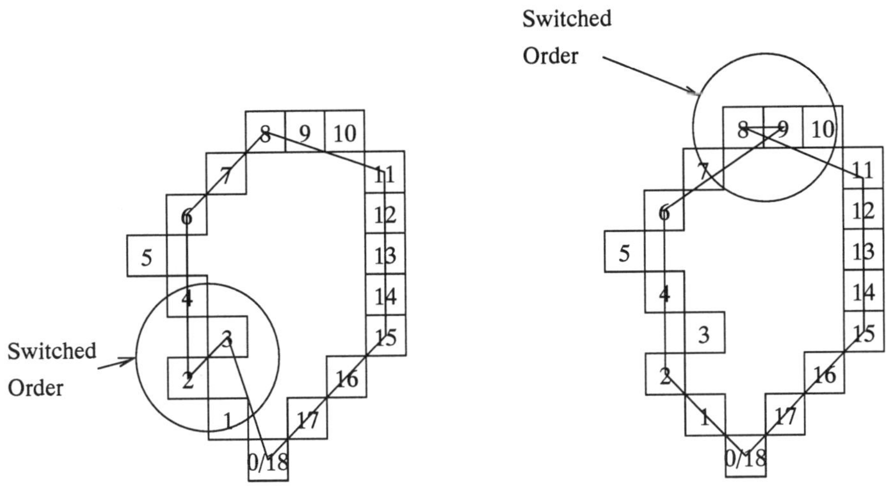

Figure 8.2 Examples of polygons with rapid changes in direction.

complexity $O(|V|^2 + |E|)$. This is a significant reduction compared to the time complexity of the exhaustive search. Recall that we defined the polygon as an ordered set for reasons which will now become apparent. We can further simplify the algorithm by observing that it is very unlikely that the optimal path will select a boundary point b_j as a vertex when the last selected vertex was b_i, where $i > j$. In general we cannot guarantee that the optimal path will not do this since the selection process depends on the vertex encoding scheme, which we have not specified yet. On the other hand, a polygon where successive vertices are not assigned to boundary points in increasing order can exhibit rapid direction changes even when the original boundary is quite smooth (see Fig. 8.2). Therefore we add the restriction that not every possible combination of (b_i, b_j) represents a valid edge but only the ones for which $i < j$. Hence the edge set E is redefined in the following way, $E = \{(b_i, b_j) \in B^2 : i < j\}$ (see Fig. 8.3). This restriction results in the fact that a given vertex set uniquely specifies the polygon. We used this fact before to derive the number of possible polygons in an exhaustive search approach.

By defining the edge set E in the above fashion, we achieve two goals simultaneously. First, the selected polygon approximation has to follow the original boundary without rapid direction changes, and second and more importantly, the resulting graph is a weighted directed acyclic graph (DAG). For a DAG,

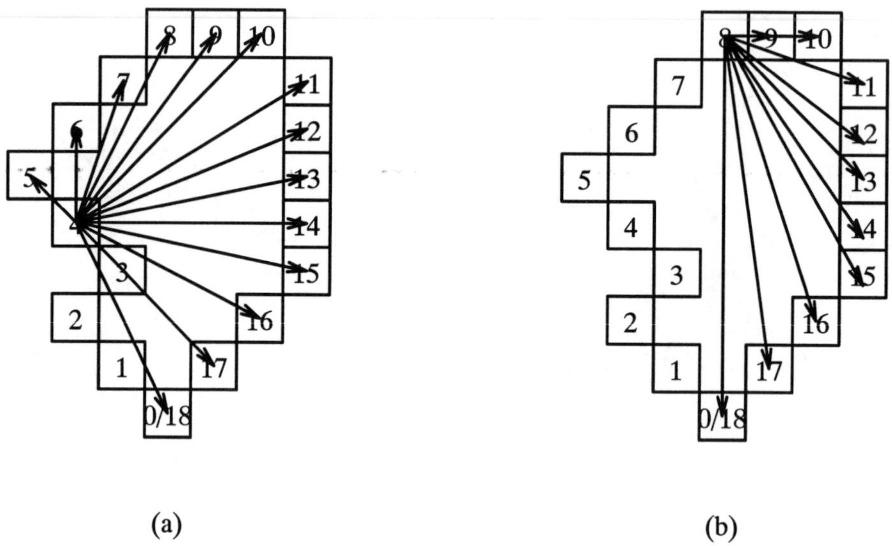

Figure 8.3 Interpretation of the boundary and the polygon approximation as a weighted directed graph. Note that the set of all edges E equals $\{(b_i, b_j) \in B^2 : i < j\}$. Two representative subsets are displayed: (a) $\{(b_4, b_j) \in B^2 : \forall\, j > 4\}$ and (b) $\{(b_8, b_j) \in B^2 : \forall\, j > 8\}$.

there exists an algorithm for finding a single-source shortest-path which is even faster than Dijkstra's algorithm. Following the notation in [34], we call this the DAG-shortest-path algorithm. The time complexity for the DAG-shortest-path algorithm is $\Theta(|V| + |E|)$, which means that the asymptotic lower bound $\Omega(|V| + |E|)$ is equal to the asymptotic upper bound $O(|V| + |E|)$.

Let $R^*(b_i)$ represent the minimum rate to reach the boundary point b_i from the source vertex $p_0 = b_0$ via a polygon approximation. Clearly $R^*(b_{N_B-1})$ is the solution to problem (8.9). Let $q(b_i)$ be a back pointer which is used to remember the optimal path. Then the proposed algorithm works as follows:

```
1)  R*(p_0) = r(p_{-1}, p_0);
2)  for i = 1, ..., N_B - 1;
3)  {
4)      R*(b_i) = ∞;
5)  }
6)  for i = 0, ..., N_B - 2;
7)  {
8)      for j = i + 1, ..., N_B - 1;
9)      {
10)         calculate edge distortion d(b_i, b_j);
11)         look up edge rate r(b_i, b_j);
12)         assign w(b_i, b_j) based on definition (8.13);
13)         if (R*(b_i) + w(b_i, b_j) < R*(b_j));
14)         {
15)             R*(b_j) = R*(b_i) + w(b_i, b_j);
16)             q(b_j) = b_i;
17)         }
18)     }
19) }
```

The optimal path $\{p_0^*, \cdots, p_{N_P-1}^*\}$ can be found by back tracking the pointers $q(b_i)$ in the following recursive fashion (by definition $p_{N_P-1}^* = b_{N_B-1}$ and $p_0^* = b_0$),

$$p_{k-1}^* = q(p_k^*), \qquad k = N_P - 1, \ldots, 2. \tag{8.14}$$

The formal proof of the correctness of the DAG-shortest-path algorithm, on which the above scheme is based, can be found in [34] and in section 3.5 we gave a graphical interpretation of the above algorithm. Here we will reason more intuitively how this approach works. In line (1) the rate for encoding the starting point of the boundary is assigned to the minimum rate of the first polygon vertex. In lines (2) to (5) the minimum rate for reaching any of the boundary points is set to infinity. The "for loop" in line (6) selects the boundary points in sequence as possible vertex points from which a polygon edge starts and the "for loop" in line (8) selects possible vertex points where the polygon edge ends. Hence these two loops select each edge in the edge set E exactly once. Therefore the lines (9) to (18) are processed for every edge. The lines (10) to (12) are used to calculate the weight of the edge, $w(b_i, b_j)$. The most important part of this algorithm is the comparison in line (13). Here we test if the new bit rate, $R^*(b_i) + w(b_i, b_j)$, to reach boundary point b_j, given that the last vertex was b_i, is smaller than the smallest bit rate used so far to reach b_j, $R^*(b_j)$. If this bit rate is indeed smaller, then it is assigned as the new smallest bit rate to reach boundary point b_j, $R^*(b_j) = R^*(b_i) + w(b_i, b_j)$. We also assign the back pointer of b_j, $q(b_j)$, to point to b_i since this is the previous vertex used to achieve $R^*(b_j)$. This algorithm leads to the optimal solution because, as stated earlier, when the rate $(R^*(b_i))$ of a vertex (b_i) is given, then the selection of the future vertices $(b_j, i < j < N_B)$ is independent of the selection of the past vertices $(b_k, 0 \leq k < i)$.

The analysis of the above algorithm shows that there are two nested loops which result in a time complexity of $\Theta(N_B^2)$. We use the number of edge distortion evaluations as measure for the time complexity, since this is the most time consuming operation. When the edge distortion is the maximum absolute distance then another loop is required because of the maximum operator in Eq. (8.3). Therefore the time complexity of the maximum absolute distance algorithm is $\Theta(N_B^3)$ with respect to the distance evaluations in Eq. (8.2).

8.3.2 The minimum distortion case

We now consider the minimum distortion case which is stated in Eq. (8.8). The goal of the proposed algorithm is to find the polygon with the smallest distortion for a given bit budget for encoding its vertices. Sometimes this is also called a rate constrained approach.

We propose an iterative solution to this problem which is based on the fact that we can solve the dual problem stated in Eq. (8.9) optimally. Consider D_{max} in Eq. (8.9) to be a variable. We derived in section 8.3.1 an algorithm which finds the polygonal approximation which results in the minimum rate for any D_{max}. We denote this optimal rate by $R^*(D_{max})$. We prove below that the rate $R^*(D_{max})$ is a non-increasing function of D_{max}, which means that $D_{max}^1 < D_{max}^2$ implies $R^*(D_{max}^1) \geq R^*(D_{max}^2)$.

Proof (by contradiction): Let polygon P_1 and rate $R^*(D_{max}^1)$ be the solutions to the minimum rate optimization problem 1. Let polygon P_2 and rate $R^*(D_{max}^2)$ be the solutions to the minimum rate optimization problem 2. Assume that $D_{max}^1 < D_{max}^2$ and $R^*(D_{max}^1) < R^*(D_{max}^2)$. Then P_1 is an admissible polygon for the optimization problem 2, since $D_{max}^1 < D_{max}^2$. Since by assumption $R^*(D_{max}^1) < R^*(D_{max}^2)$, P_1 is a better solution than P_2, which is a contradiction since we showed that the selection algorithm employed to find P_2 is optimal. Hence $D_{max}^1 < D_{max}^2$ implies $R^*(D_{max}^1) \geq R^*(D_{max}^2)$. ■

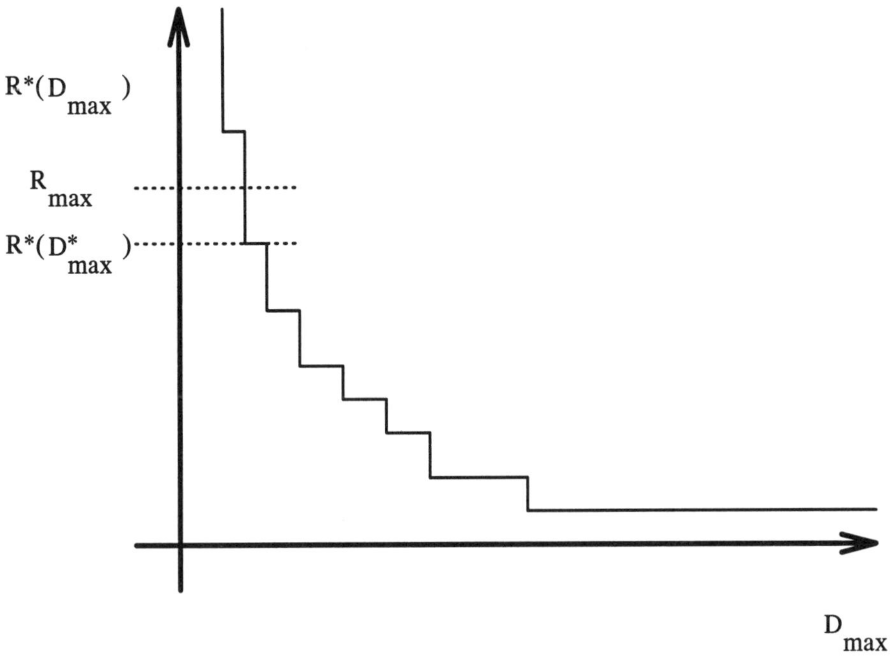

Figure 8.4 The $R^*(D_{max})$ function, which is a non-increasing function exhibiting a staircase characteristic. The selected R_{max} falls onto a discontinuity and therefore the optimal solution is of the form $R^*(D^*_{max}) < R_{max}$, instead of $R^*(D^*_{max}) = R_{max}$.

Having shown that $R^*(D_{max})$ is a non-increasing function, we can use bisection [59] to find the optimal D^*_{max} such that $R^*(D^*_{max}) = R_{max}$. Since this is a discrete optimization problem, the function $R^*(D_{max})$ is not continuous and exhibits a staircase characteristic (see Fig. 8.4). This implies that there might not exist a D^*_{max} such that $R^*(D^*_{max}) = R_{max}$. In that case the proposed algorithm will still find the optimal solution, which is of the form $R^*(D^*_{max}) < R_{max}$, but only after an infinite number of iterations (see section 4.3.3). Therefore if we have not found a D_{max} such that $R^*(D_{max}) = R_{max}$ after a given maximum number of iterations, we terminate the algorithm.

8.4 DISTORTION MEASURES BASED ON THE SUMMATION OPERATOR

In this section we introduce two algorithms to solve the problems stated in Eqs. (8.8) and (8.9) for class two distortion measures, such as the mean squared distance. Both presented algorithms are symmetric in the rate and the distortion and hence the same technique can be employed for the minimum distortion case (Eq. (8.8)) and the minimum rate case (Eq. (8.9)). We will therefore only solve the minimum distortion case and the minimum rate case can be solved be applying the following relabeling to the function names,

$$D(p_0, \ldots, p_{N_P-1}) \leftarrow R(p_0, \ldots, p_{N_P-1}), \tag{8.15}$$
$$R(p_0, \ldots, p_{N_P-1}) \leftarrow D(p_0, \ldots, p_{N_P-1}). \tag{8.16}$$

The first algorithm we propose is based on the Lagrangian multiplier method. Like every Lagrangian-based approach the resulting solutions belong to the convex hull of the operational rate distortion function. For cases where the Lagrangian bound is not tight enough, we propose a tree pruning-based scheme, which can find all optimal solutions.

8.4.1 Lagrangian multiplier approach

In this section we derive a solution to problem (8.8) which is based on the Lagrangian multiplier method we presented in section 3.3 and the shortest path algorithm presented in section 8.3.1. In this application we will use the Lagrangian multiplier method to relax the constraint so that the relaxed problem can be solved using the shortest path algorithm proposed in section 8.3.1.

We first define the Lagrangian cost function

$$J_\lambda(p_0, \ldots, p_{N_P-1}) = D(p_0, \ldots, p_{N_P-1}) + \lambda * R(p_0, \ldots, p_{N_P-1}), \tag{8.17}$$

where λ is called the Lagrangian multiplier. We showed in section 3.3 that if there is a λ^* such that

$$\{p_0^*, \ldots, p_{N_P-1}^*\} = \arg \min_{p_0, \ldots, p_{N_P-1}} J_{\lambda^*}(p_0, \ldots, p_{N_P-1}) \tag{8.18}$$

leads to $R(p_0, \ldots, p_{N_P-1}) = R_{max}$, then $\{p_0^*, \ldots, p_{N_P-1}^*\}$ is also an optimal solution to (8.8). It is well known that when λ sweeps from zero to infinity, the solution to problem (8.18) traces out the convex hull of the operational rate distortion function, which is a non-increasing function. Hence bisection [59] or the fast convex search we presented in section 4.2 can be used to find λ^*. Therefore, if we can find the optimal solution to the unconstrained problem (8.18), then we can find the optimal λ^* and the convex hull approximation to the constrained problem of Eq. (8.8).

The key observation for deriving an efficient search for the polygon which minimizes the unconstrained problem (8.18) is based on the fact that given a certain vertex of a polygon (p_{k-1}) and the Lagrangian cost function which results by coding the polygon up to and including this vertex ($J_{k-1}(p_{k-1}) = D_{k-1}(p_{k-1}) + \lambda * R_{k-1}(p_{k-1})$), the selection of the next vertex p_k is independent of the selection of the previous vertices $p_0, \ldots, p_{k-2}$. This is true since the rate and the distortion can be expressed recursively as functions of the segment rates $r(p_{k-1}, p_k)$ and segment distortions $d(p_{k-1}, p_k)$,

$$R_k(p_k) = R_{k-1}(p_{k-1}) + r(p_{k-1}, p_k), \tag{8.19}$$

and

$$D_k(p_k) = D_{k-1}(p_{k-1}) + d(p_{k-1}, p_k). \tag{8.20}$$

These recursions need to be initialized by setting $R_{-1}(p_{-1})$ and $D_{-1}(p_{-1})$ equal to zero. Since the rate and the distortion can be calculated with a first order recursion, the Lagrangian cost function can also be calculated recursively,

$$J_k(p_k) = J_{k-1}(p_{k-1}) + \{d(p_{k-1}, p_k) + \lambda * r(p_{k-1}, p_k)\}. \tag{8.21}$$

Clearly $R_{N_P-1}(p_{N_P-1})$ is equal to $R(p_0, \ldots, p_{N_P-1})$, the rate for the entire polygon, $D_{N_P-1}(p_{N_P-1})$ is equal to $D(p_0, \ldots, p_{N_P-1})$ the distortion for the entire polygon and $J_{N_P-1}(p_{N_P-1})$ is equal to $J(p_0, \ldots, p_{N_P-1})$, the Lagrangian cost function for the entire polygon.

In section 8.3.1 we have shown that an optimization problem which has the above described structure can be solved optimally by a DAG-shortest-path algorithm. To be able to employ the previously proposed shortest path algorithm, we have to redefine the weight function $w(u, v) : E \rightarrow \mathcal{R}$,

$$w(u, v) = d(u, v) + \lambda * r(u, v). \tag{8.22}$$

Since the shortest path algorithm results in the polygon which minimizes the following sum,

$$\sum_{k=0}^{N_P-1} w(p_{k-1}, p_k), \tag{8.23}$$

this polygon is the optimal solution to the relaxed problem of Eq. (8.18).

Clearly, the time complexity of the Lagrangian approach for a fixed λ is the same as for the shortest path algorithm. The shortest path algorithm is invoked several times by the bisection algorithm to find the optimal λ^* and hence the time complexity is a function of the number of required iterations. As pointed out before, the Lagrangian approach finds optimal solutions which belong to the convex hull of the operational rate distortion curve. Clearly there are other optimal solutions which are above the convex hull. In the next section we present a tree pruning algorithm which finds all optimal solutions.

8.4.2 Pruning Approach

As before, for this algorithm we make use of the fact that when the current vertex is selected and we know the rate and the distortion used to encode the polygon up to and including this vertex, the previous vertices do not influence the selection of the future vertices. For a given boundary point (b_i) under consideration to be chosen as a polygon vertex, every previous boundary point ($b_k, 0 \leq k < i$) could have been the last vertex used for the polygon approximation. Therefore, the rate and distortion for every previous boundary point are calculated under the assumption that the previous boundary point was used as the previous vertex. These calculations lead to a set of nodes, each representing the hypothesis that the current boundary point is a vertex but with different boundary points as previous vertices.

We introduce a pruning procedure to reduce the number of nodes for each boundary point. If there are two nodes j and i such that $D(j) \geq D(i)$ and $R(j) \geq R(i)$, where $D(n)$ is the distortion and $R(n)$ the rate up to and including node n, then it is clear that node j cannot belong to the optimal solution. This is because node i has a lower distortion and a lower rate than node j, but both represent the same boundary point as the last selected vertex. Hence node j is pruned from the decision tree. Since the pruned nodes need not be considered in the future of the optimization process, the more nodes pruned, the faster the algorithm becomes. A straightforward approach to pruning has a quadratic time complexity in the number of nodes to be pruned NP. Since NP depends on previous pruning results, the time complexity of the entire approach depends on the boundary, the distortion measure and the vertex encoding scheme. As with most integer programming algorithms, one can construct an example where the pruning scheme fails completely, which results in an exponential time complexity. Note that this exponential time complexity is still better than the time complexity of the exhaustive search, since we use the fact that the future is independent of the past. In general though, the pruning is extremely efficient in cutting down the complexity of the algorithm and in fact, this scheme and the previously discussed Lagrangian approach take about the same amount of time for the experimental results we will present in section 8.9. If the complete set of optimal rate distortion points is not of interest, but only the problem of Eq. (8.8) needs to be solved, additional pruning can be achieved by removing all nodes which contain a rate higher than R_{max}. This leads effectively to all optimal rate distortion points below and including the line $R(D) = R_{max}$.

Each of the remaining nodes represents a polygon which has the current boundary point as its last vertex, but with different rate-distortion characteristics. In other words, the remaining nodes represent the set of all optimal solutions for the encoding of the boundary up to and including the current boundary point. These nodes make up the admissible nodes for this boundary point, when this boundary point is considered as a previous vertex in the future of the optimization process.

Figure 8.5 shows a simple example of the algorithm. In the upper left corner is the boundary which must be encoded. The adjacent pixels are labeled 0, 1, 2, 3 and 4 and they simply form a square of side length 1. Note that point 4 is the same as point 0, and therefore it does not need to be transmitted, but still a distortion occurs between the last vertex of the polygon and point 4 and it is therefore included in the closed boundary. Figure 8.5 shows the complete decision tree. This tree reflects the fact that given the boundary point used as the previous vertex, and the rate and the distortion for encoding the polygon up to and including that vertex, the selection of future vertices is independent of the selection made for previous vertices. In Fig. 8.5 the boundary point index, the rate and the distortion are indicated in the following fashion: "index/rate/distortion". In this example, the sum of the squared distance between the boundary and the polygon is used as the distortion measure and the vertex encoding scheme proposed in section 8.8 is employed, (4 bits are required for each transition in this example).

There are two possible transitions from a given node. The upward transition, which indicates that this node is used as a previous vertex and the downward transition which indicates that this node is not used as a previous vertex. The downward transitions carry a weight of zero (rate = 0, distortion = 0), but the upward transitions result in the addition of r(previous vertex, current vertex) to the rate and d(previous vertex, current vertex) to the distortion. The epochs, which correspond to the boundary points, are indicated at the bottom of the tree. The boxes indicate the new nodes per boundary point and therefore the pruning procedure is only applied to those nodes.

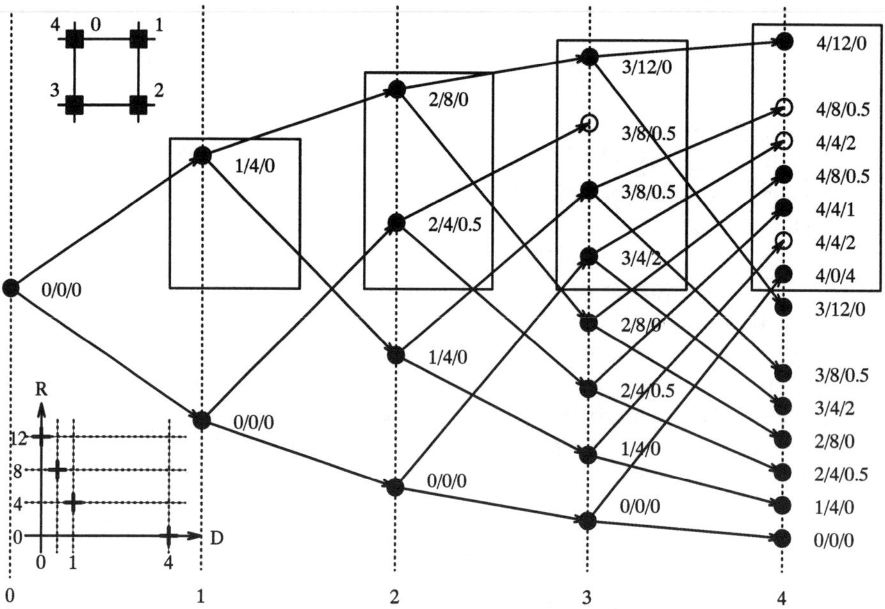

Figure 8.5 Pruned decision tree for the encoding of a boundary. The nodes are labeled as follows: "index/rate/distortion".

Consider the box at epoch 3. There are two nodes, (both with description 3/8/0.5) which require the same rate and lead to the same distortion to reach boundary point 3. Therefore one of the two can be pruned (indicated by an empty circle) since both will lead to the same collection of future paths. By pruning a node, the collection of future paths gets reduced. Clearly, the more nodes that can be pruned, the faster the algorithm is.

At the last epoch, which corresponds to boundary point 4, 3 nodes can be pruned and only 4 final nodes remain which represent the 4 optimal solutions to the boundary approximation. These 4 optimal solutions are also displayed as an operational rate distortion function in the lower left corner of Fig. 8.5. The path (0,1,2,3,4) which leads to 12 bits and no distortion is the highest quality approximation which is basically the chain code of the original boundary. The path (0,1,2,2,4) approximates the box by a triangle which requires 8 bits and leads to a distortion of 0.5. The path (0,0,2,2,4) approximates the box by a diagonal line which requires 4 bits and leads to a distortion of 1. Finally the path (0,0,0,0,4) does not require any bits, since it approximates the box by its starting point, but it leads to a maximum distortion of 4.

It is interesting to note that this pruning scheme can be easily modified to work with class one distortion measures, since the additivity of the class two distortion measures has not been used in the derivation of this scheme. The only fact employed is the independence of the future from the past, which is also present for class one distortion measures. In general though, the schemes presented for class one distortion measures are faster than the pruning approach.

8.5 INCLUDING SECONDARY OBJECTIVES

In this section we propose two methods for including a secondary objective into the optimization problem, which is used to select among different optimal polygons. For example, there might be two different polygons which use the same number of bits and result in the same maximum error. Hence both polygons are optimal for the minimum maximum distortion criterion and we can select either one to be the approximation of the boundary. In such a case, it is natural to pick the one which, for example, results in a smaller mean squared distance. This idea can be generalized in the sense that for distortion measures of class one, we would like to use a distortion measure of class two to select the final approximation among the optimal polygons. Vice versa, for distortion measures of class two, we would like to use a distortion measure of class one to select the final approximation among the optimal polygons. A straightforward approach based on this idea is to run the algorithm and to select the best polygon among the optimal polygons available at the last node. Unfortunately, this approach has the flaw that during the process of finding these optimal polygons, ties had to be broken in the case where two or more partial polygons where equally good up to a given node. To truly find the best polygon, the one which is optimal based on the primary distortion measure and, of all optimal polygons, the one which results in the smallest secondary distortion, each tie in the algorithm has to be broken using the secondary distortion measure. This is the basic idea the following two schemes are based on.

8.5.1 Distortion measures based on the maximum operator

We have shown in section 8.3 that for distortion measures of class one the optimal solution to the minimum rate case can be used to solve the minimum distortion case. Hence we concentrate on the minimum rate case, which is stated in Eq. (8.9). The basic idea behind the minimum rate case is that we only permit edges which result in an edge distortion smaller than or equal to the maximum distortion. The weight of each of these edges is then set to the rate required to encode the edge, and hence the DAG shortest path algorithm will find the polygon with the smallest rate which does not violate the maximum polygon distortion. This concept can be extended to include the above stated idea of a secondary objective which is used to break a tie when there are more than one optimal polygon. We would like to use a class two distortion measure to break the tie, such as the mean squared error. Recall that class two distortion measures are polygon distortion measures which are the sum of the edge distortion measures (see Eq. (8.6)). We therefore propose to change the weight function $w(p_{k-1}, p_k)$ from Eq. (8.10), in the following way,

$$w(p_{k-1}, p_k) = \begin{cases} \infty & : \; d(p_{k-1}, p_k) > D_{max} \\ r(p_{k-1}, p_k) + \beta * \tilde{d}(p_{k-1}, p_k) & : \; d(p_{k-1}, p_k) \leq D_{max} \end{cases} . \tag{8.24}$$

where $\tilde{d}(p_{k-1}, p_k)$ is the edge distortion measure which is used for the class two secondary polygon distortion measure, and β is a positive real number. The proposed DAG shortest path algorithm finds the polygon which minimizes the following sum,

$$\sum_{k=0}^{N_P-1} \left(r(p_{k-1}, p_k) + \beta * \tilde{d}(p_{k-1}, p_k) \right) . \tag{8.25}$$

Clearly if β is zero, then nothing changed and hence there might be several polygons which result in the same smallest rate.

Let the smallest possible difference between the rate for the optimal polygons and the rate for the next best polygon be denoted by ΔR, which for a Huffman code based scheme is one bit. Let the largest possible distortion of the secondary class two polygon distortion measure be denoted by $\tilde{D}_{max}$. We would like to select β such that it will identify a winner among the optimal polygons, but it should be impossible for any other polygon to outperform the optimal polygons. This can be achieved by selecting β as follows,

$$\frac{\Delta R}{\tilde{D}_{max}} > \beta > 0. \tag{8.26}$$

By selecting β as proposed above, even in the worst case, when the best rate and the second best rate are only ΔR apart, the secondary polygon distortion of the best polygon is the maximum secondary polygon distortion $\tilde{D}_{max}$, and the secondary polygon distortion of the second best polygon is zero, the sum calculated in Eq. (8.25) for the best polygon is still smaller than for the second best polygon. Hence the second best polygon will never be selected as the optimal polygon, and the secondary distortion measure is only used to find a winner among the optimal polygons.

8.5.2 Distortion measures based on the summation operator

We have introduced two different approaches for finding an optimal polygon for class two distortion measures, the Lagrangian approach and the pruning approach. As in the previous section, we would like to use a class one distortion measure to break the ties in the proposed algorithms for class two distortion measures. In the previous section we were able to include the secondary distortion measure very elegantly into the existing algorithm, but in this section, more modifications of the original algorithms are necessary. Since class one secondary distortion measures are based on the maximum operator, we cannot simply include their cost into the optimization procedure. Nevertheless, the knowledge of the maximum distortion up to and including the current vertex, makes the future of the maximum distortion independent of the past. We will use this concept to incorporate the secondary distortion measure in the two proposed algorithms for class two polygon distortion measures.

Lagrangian approach

As we pointed out in section 8.4.1 the Lagrangian cost function exhibits the same first order state dynamics as the secondary class one polygon distortion measure. Hence the Lagrangian cost function up to and including a certain vertex b_i, $J^*(b_i)$, and the maximum secondary edge distortion up to and including the vertex under consideration, $\tilde{D}^*(b_i)$, have to be stored within the vertex. In the case that there is a tie between the Lagrangian cost function up to and including the vertex under consideration, the partial polygon which results in a smaller secondary distortion is selected. For a given Lagrangian multiplier λ, this results in the following DAG shortest path algorithm, which is almost identical to the one presented in section 8.3, except for the changes mentioned above.

1) $J^*(p_0) = r(p_{-1}, p_0) + \lambda * d(p_{-1}, p_0);$
2) for $i = 1, \ldots, N_B - 1;$
3) {
4) $\quad J^*(b_i) = \infty;$
5) $\quad D^*(b_i) = \infty;$
6) }
7) for $i = 0, \ldots, N_B - 2;$
8) {
9) $\quad$ for $j = i + 1, \ldots, N_B - 1;$
10) $\quad$ {
11) $\qquad$ calculate primary edge distortion $d(b_i, b_j);$
12) $\qquad$ calculate secondary edge distortion $\tilde{d}(b_i, b_j);$
13) $\qquad$ look up edge rate $r(b_i, b_j);$
14) $\qquad w(b_i, b_j) = r(b_i, b_j) + \lambda * d(b_i, b_j);$
15) $\qquad$ if $\left\{ \left[J^*(b_i) + w(b_i, b_j) < J^*(b_j) \right] \text{ or } \left[\left(J^*(b_i) + w(b_i, b_j) = J^*(b_j) \right) \text{ and } \left(\max\{D^*(b_i), \tilde{d}(b_i, b_j)\} < D^*(b_j) \right) \right] \right\}$
16) $\qquad$ {
17) $\qquad\quad J^*(b_j) = J^*(b_i) + w(b_i, b_j);$
18) $\qquad\quad D^*(b_j) = \max\{D^*(b_i), \tilde{d}(b_i, b_j)\};$
19) $\qquad\quad q(b_j) = b_i;$
20) $\qquad$ }
21) $\quad$ }
22) }

The main change necessary to use a secondary distortion measure to break ties is in the "if" statement in line (15). As before, the lines (16) to (20) are executed if the Lagrangian distortion is smaller than the one currently associated with the vertex. In the case when there is a tie, lines (16) to (20) are only executed if the secondary distortion measure is smaller than the one currently associated with the vertex.

Pruning approach

The main difference between the Lagrangian approach and the pruning approach is the fact that the pruning approach can also find solutions which do not belong to the convex hull of the operational rate distortion function. The approach we take to incorporate the secondary distortion measure is very similar to the one we used for the Lagrangian approach. Again we make use of the first order state dynamics inherent in class one distortion measures. The main idea behind the pruning approach is to create all the possible rate distortion pairs for a given vertex and then prune the ones which do not belong to the operational rate distortion curve. Hence each node in the pruning approach contains its index, its rate and its distortion. We propose to add the maximum secondary distortion up to and including this vertex $\tilde{D}^*(n)$ to the information stored in a node. During pruning, if we encounter the case that there are two or more nodes which have the same rate and the same distortion, we will retain the one with the smallest secondary distortion. By changing the pruning strategy in this fashion, the resulting optimal solution will be the optimal solution with the smallest secondary distortion.

8.6 EXTENSION OF THE ADMISSIBLE VERTEX SET TO OFF-BOUNDARY PIXELS

In this section we relax the restriction, stated in section 8.2, that the admissible vertices for the polygon must belong to the original boundary. As we pointed out before, the main advantage of this restriction is that the polygon distortion can be expressed as the maximum or the sum over the edge distortions (see Eqs. (8.3) and (8.4)). We used this fact to find the optimal polygon using a fast DAG algorithm. The main disadvantage of this constraint comes from the fact that one can easily construct an example where for a given maximum polygon distortion, the polygon with the smallest bit rate uses vertices which do not fall onto boundary points. The problem with using vertices which do not belong to the boundary is that for a given polygon edge no direct correspondence exits between the edge and a subset of boundary points. Hence the polygon distortion cannot be formulated as the maximum or the sum of the edge distortions since the edge distortions are not defined.

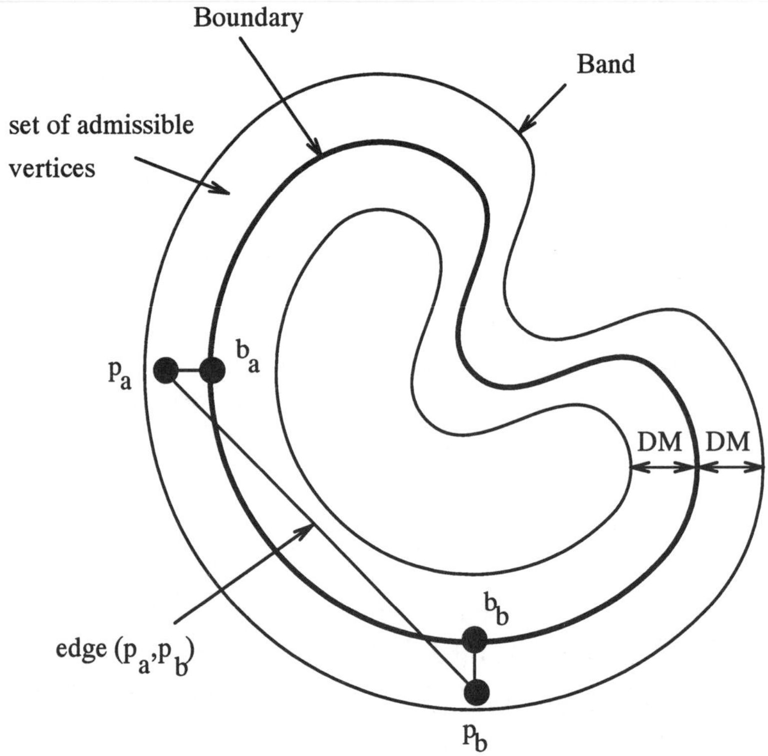

Figure 8.6 The "band" concept.

8.6.1 Definition of the admissible vertex set

From a theoretical point of view, the set of admissible vertices should contain all the pixels in the image plane. On the other hand, the DAG shortest path algorithm has a time complexity which is quadratic in the number of admissible vertices and hence we would like to keep that number as small as possible, without sacrificing coding efficiency. In the proposed approach, the set of all admissible vertices is defined as all the pixels which are within a given maximum distance DM from a boundary point (see Fig. 8.6). Hence the set of admissible vertices consist of a "band" of width $2 * DM$ around the original boundary. Clearly, the bit rate for the optimal polygon is a non-increasing function of the size of the proposed band. On the other hand, the time complexity of the DAG shortest path algorithm is quadratic in the number of admissible vertices and the larger the band, the bigger the set of admissible vertices.

8.6.2 Definition of the edge distortion measure

The basic idea behind adapting the DAG shortest path algorithm for the case where off-boundary vertices are permissible, is to re-define the edge distortion measure. Since the DAG shortest path algorithm requires edge distortions, we would like to define edge distortions in such a way that the maximum or the sum of these edge distortions is smaller than or equal to the polygon distortion. Hence the optimal polygon found using these edge distortions will always satisfy the maximum polygon distortion and we can still employ the fast graph algorithm for the optimization procedure.

We therefore propose a new edge distortion definition such that the resulting optimal polygon will always satisfy the maximum polygon distortion and the new distortion measure leads to the desired independence between edges. Every admissible vertex (for example p_b in Fig. 8.6) is associated with the closest boundary point (which is b_b for p_b in Fig. 8.6). The distortion of two admissible vertices (the edge distortion $d(p_a, p_b)$) is then defined using the polygon edge ((p_a, p_b) in Fig. 8.6) and the partial boundary ($\{b_a = p_a, \ldots, b_b = p_b\}$ in Fig. 8.6). Again, this edge distortion can be the maximum distance, as described in Eq. (8.3), the mean squared distance as described in Eq. (8.4), or any other distortion measure such as the absolute area or the number of error pixels. This definition of the edge distortion leads to the required independence between the edge distortions and hence the fast DAG algorithm can be used to find the optimal polygon, where optimality is defined with respect to this edge distortion measure.

8.6.3 Ordering of the admissible vertices

As discussed in section 8.3.1, the set of all admissible vertices should be an ordered set, otherwise Dijkstra's algorithm, which is slower than the DAG algorithm, needs to be employed. The ordering of the admissible vertices is straightforward in the case when they belong to the boundary. We just use the order of the boundary set and impose it on the vertex set. Since the vertex set is now bigger than the boundary set, this ordering is not as obvious anymore.

The original motivation for the ordering comes from the observation that we should not allow for backwards transitions, such as the ones displayed in Fig. 8.2. The translation of this idea to the band concept is that we would like to order the admissible vertices by assigning them to the nearest boundary point and then the order of the boundary points implies an order for the assigned admissible vertices. Since we restrict the vertices to fall on pixels, the closest boundary point of a given admissible vertex is not necessarily unique. We propose the following procedure to define the set of admissible vertices V and to assign the vertices to boundary points.

```
1)  V=B;
2)  i=1;
3)  while ( d(v0,vi) ≤ DM )
4)  {
5)      for j = 1,...,N_B − 2;
6)      {
7)          if ( {bj + vi} ∉ V )
8)          {
9)              V = V ∪ {bj + vi};
10)             assign vertex bj+vi to boundary point bj;
11)         }
12)     }
13)     i=i+1;
14) }
```

The v_i in the above algorithm are offset vectors, which are displayed in Fig. 8.7. In Fig. 8.7, the vector v_j starts at 0 and ends in j. Note that the distance $d(v_0, v_j)$ is a non-decreasing function of j and it is displayed in Fig. 8.8. The basic idea behind the proposed algorithm is to draw spirals around the boundary points, where the spirals are drawn in discrete increments of pixels as can be seen in Fig. 8.8. After a spiral is advanced by one increment, the new pixel of the spiral is an admissible vertex and it is associated with the boundary point in the center of the spiral. Then this procedure is repeated for the next boundary point until there are no more boundary points. Then the procedure starts with the first boundary point again until the distance is bigger than DM.

In the above algorithmic description of this scheme, the inner loop (lines 4 to 10) tries to assign one admissible vertex b_j+v_i to each boundary point b_j. This loop

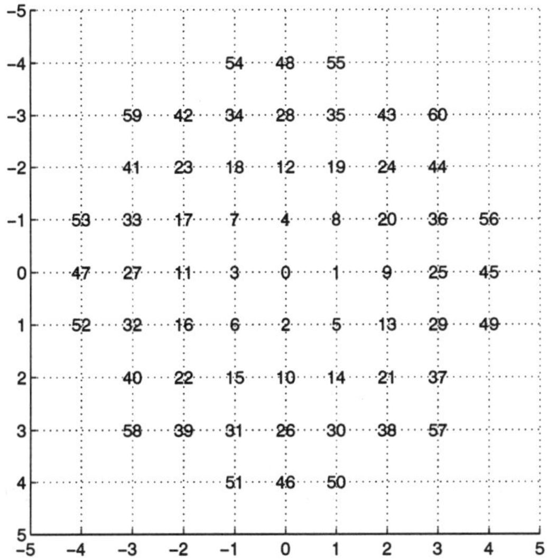

Figure 8.7 Vector increments for the ordering of the admissible vertices. Note that vector v_j starts at 0 and ends at j.

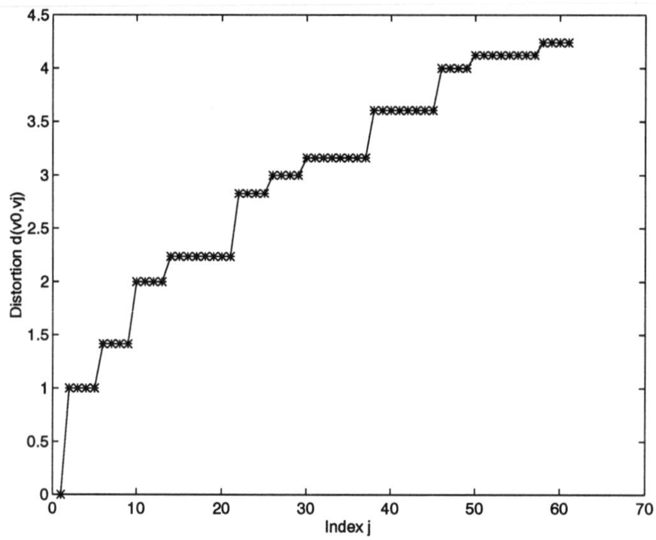

Figure 8.8 Distance as a function of the index.

is repeated for the next vector increment v_{i+1} as long as the admissible vertices are within the band defined by the maximum distance DM. In Fig. 8.9 different stages of the algorithm are shown. In this example, the maximum distance DM is set to one pixel and therefore only the vectors v_1, v_2, v_3 and v_4 are used in the algorithm, since all other vectors result in a higher distortion as can be seen from Fig. 8.8. Fig. 8.9a shows the admissible vertices and the assignment of these vertices to the boundary points after the vector increment v_1 has been used for all boundary points. Fig. 8.9b shows the admissible vertices and the assignment of these vertices to the boundary points after the vector increment v_2 has been used for all boundary points. Fig. 8.9c shows the admissible vertices and the assignment of these vertices to the boundary points after the vector increment v_3 has been used for all boundary points and Fig. 8.9d shows the admissible vertices and the assignment of these vertices to the boundary points after the vector increment v_3 has been used for all boundary points, which is the final solution.

The proposed ordering algorithm results in a set of admissible vertices, which forms a band around the original boundary of width $2 * DM$, and an ordering of the admissible vertices which enables us to use the fast DAG shortest path

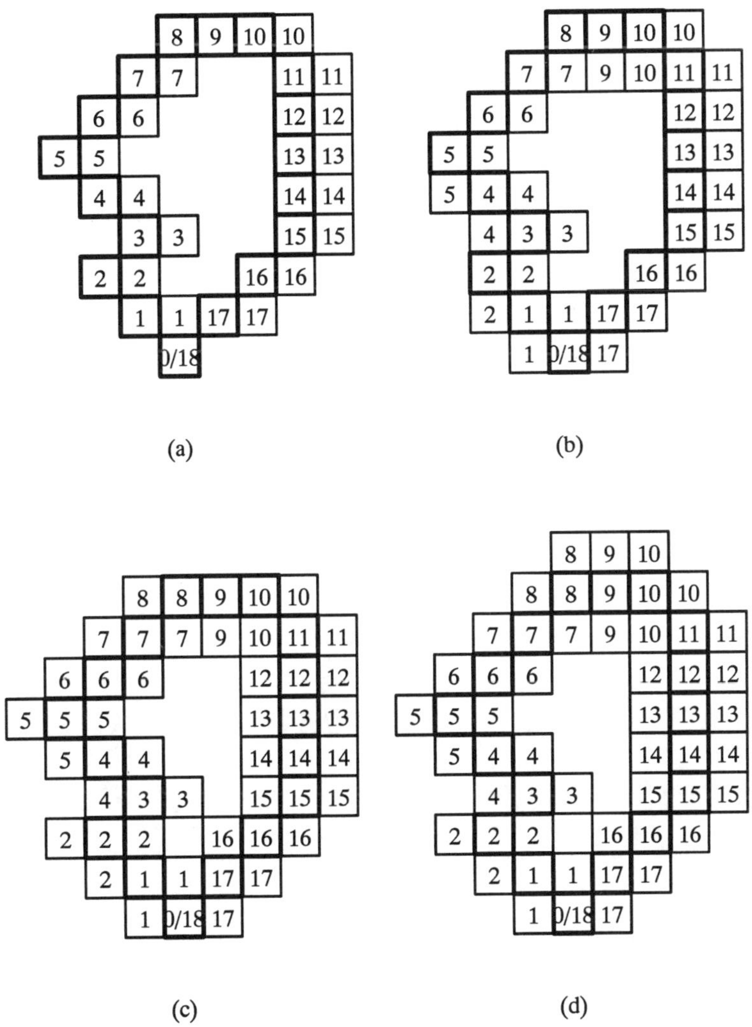

Figure 8.9 Figure a, b, c, and d show the final result after the inner loop has been finished for v_1, v_2, v_3 and v_4, which corresponds to a maximum distance DM of one pixel. The pixels with a bold border indicate the original boundary and the other pixels are the admissible vertices, where the number of the associated boundary point is written inside the admissible vertex.

algorithm instead of Dijkstra's algorithm. The imposed ordering is such that we only admit directed edges where the number of the "from" vertex is lower than the number of the "to" vertex. Again, this restriction will improve the speed of the algorithm and it also forces the optimal polygon to follow the original boundary. There are different ways that the order of the vertices with the same number can be defined, and for our purposes we consider them of equal order and hence no edge can start and end at vertices which belong to the same boundary point. The proposed ordering algorithm has several advantages: (1) It defines the set of admissible vertices, so that the resulting set forms a band of width $2 * DM$ around the original boundary. (2) Each admissible vertex is automatically assigned to a closest boundary point and (3) the set of admissible vertices is ordered along the original boundary.

Note that using the above definition of the edge distortion and the admissible set of vertices, the algorithms introduced for the case when the vertices belonged to the original boundary can still be applied. Clearly since the time complexities of all the proposed algorithms are quadratic in the number of admissible vertices, we should select the size of the band DM to be as small as possible. In the experiment we present in section 8.9, DM was set to one pixel, which for the employed vertex encoding scheme and the boundaries we tested, was a good compromise between speed and coding performance.

8.7 MULTIPLE BOUNDARY ENCODING

In this section we extend the results of the previous sections for the encoding of multiple boundaries. Assume that M different boundaries have to be encoded. We will adopt the convention that a subscript indicates which boundary is addressed, i.e., B_3 is the third boundary, P_4 is the polygon used to approximate the fourth boundary, etc. Then the minimum rate optimization problem can be stated as follows,

$$\min_{P_0,\ldots,P_{M-1}} R(P_0,\ldots,P_{M-1}), \qquad \text{subject to:} \qquad D(P_0,\ldots,P_{M-1}) \leq D_{max}, \tag{8.27}$$

whereas the minimum distortion problem is of the following form,

$$\min_{P_0,\ldots,P_{M-1}} D(P_0,\ldots,P_{M-1}), \qquad \text{subject to:} \qquad R(P_0,\ldots,P_{M-1}) \leq R_{max}. \tag{8.28}$$

The total rate $R(P_0, \ldots, P_{M-1})$ in the above formulation is defined as

$$R(P_0, \ldots, P_{M-1}) = \sum_{i=0}^{M-1} R_i(p_{i,0}, \ldots, p_{i,N_{P_i}-1}). \tag{8.29}$$

The total distortion measure $D(P_0, \ldots, P_{M-1})$ is defined for class one distortion measures by,

$$D(P_0, \ldots, P_{M-1}) = \max_{i \in [0, \ldots, M-1]} D_i(p_{i,0}, \ldots, p_{i,N_{P_i}-1}), \tag{8.30}$$

and for class two distortion measures by

$$D(P_0, \ldots, P_{M-1}) = \sum_{i=0}^{M-1} D_i(p_{i,0}, \ldots, p_{i,N_{P_i}-1}). \tag{8.31}$$

As we have seen in the previous sections, the two classes of distortion measures require different algorithms. This is also true for the encoding of multiple boundaries.

8.7.1 Distortion measures based on the maximum operator

The minimum rate case

Since the total rate for the encoding of all boundaries, $R(P_0, \ldots, P_{M-1})$, is the sum of the rates, $R_i(p_{i,0}, \ldots, p_{i,N_{P_i}-1})$, for the encoding of the individual boundaries, and the encoding of the individual boundaries is accomplished independently, the minimum total rate is equal to the sum of the minimum individual rates, where the search for the minimum individual rates is also constrained by the maximum distortion D_{max}. Therefore the following optimization problem is identical to the one in Eq. (8.27),

$$\begin{aligned} &\text{for } i = 0, \ldots, M-1, \\ &\min_{p_{i,0},\ldots,p_{i,N_{P_i}-1}} \quad R_i(p_{i,0}, \ldots, p_{i,N_{P_i}-1}), \\ &\quad \text{subject to:} \quad D_i(p_{i,0}, \ldots, p_{i,N_{P_i}-1}) \le D_{max}, \end{aligned} \tag{8.32}$$

which shows that the optimal solution to problem (8.27) can be found by solving the optimization problems for the different boundaries independently using the algorithm developed in section 8.3.1.

The minimum distortion case

We now consider the minimum distortion case of Eq. (8.28). As in section 8.3.2 we use the fact that we can solve the minimum rate problem optimally, to solve the minimum distortion problem by an iterative scheme. By defining $R^*(D_{max})$ of section 8.3.2 as the minimum total rate needed to encode the M given boundaries with a maximum error of D_{max}, the derivation in section 8.3.2 still applies. Hence the resulting algorithm is a bisection search and at each iteration the optimization problem of Eq. (8.27) is solved optimally using the above proposed scheme.

8.7.2 Distortion measures based on the summation operator

For class two distortion measures, we have introduced two different algorithms, the Lagrangian multiplier approach and the tree-pruning scheme. As was pointed out in section 8.4, these algorithms are symmetric in the rate and the distortion and therefore we will only discuss the minimum distortion case of Eq. (8.28).

Lagrangian multiplier approach

We define the total Lagrangian cost function $J_\lambda(P_0, \ldots, P_{M-1})$ as follows,

$$J_\lambda(P_0, \ldots, P_{M-1}) = \sum_{i=0}^{M-1} \left\{ D_i(p_{i,0}, \ldots, p_{i,N_{P_i}-1}) + \lambda * R_i(p_{i,0}, \ldots, p_{i,N_{P_i}-1}) \right\}. \tag{8.33}$$

According to section 8.4.1, if we can find the global minimum of the total Lagrangian cost function with respect to the approximation polygons,

$$\begin{aligned} &\min_{P_0,\ldots,P_{M-1}} J_\lambda(P_0, \ldots, P_{M-1}) \\ &= \min_{P_0,\ldots,P_{M-1}} \left\{ \sum_{i=0}^{M-1} D_i(p_{i,0}, \ldots, p_{i,N_{P_i}-1}) + \lambda R_i(p_{i,0}, \ldots, p_{i,N_{P_i}-1}) \right\} \end{aligned} \tag{8.34}$$

then we can use an iterative search to solve the constrained problem of Eq. (8.28).

Since the different boundaries are independently encoded, the minimum of the total Lagrangian cost function can be found by minimizing each of the individual boundaries separately. In other words, this problem reduces to the ones studied in [191]. Hence the following minimization is equivalent to the one in Eq. (8.33),

$$\min_{P_0,\ldots,P_{M-1}} J_\lambda(P_0,\ldots,P_{M-1})$$
$$=\sum_{i=0}^{M-1} \min_{p_{i,0},\ldots,p_{i,N_{P_i}-1}} \{D_i(p_{i,0},\ldots,p_{i,N_{P_i}-1})+\lambda R_i(p_{i,0},\ldots,p_{i,N_{P_i}-1})\} \quad (8.35)$$

Therefore the multiple boundary encoding problem can be solved with the Lagrangian multiplier method using Eq. (8.34), which states that the global minimum of the total Lagrangian cost function is the sum of the global minima of the Lagrangian cost functions for each boundary. Since we derived in section 8.4.1 an algorithm to find the minimum of a boundary Lagrangian cost function, we can find the minimum of Eq. (8.34), and hence we can encode multiple objects using the Lagrangian relaxation scheme.

Pruning approach

The tree-pruning approach proposed in section 8.4.2 results in the operational rate distortion function (ORDF) for a given boundary. Again, since the boundaries are encoded independently, we first run the pruning algorithm for each of the M boundaries. This results in M different ORDFs. It is interesting to notice that the optimal bit allocation among independent quantizers (characterized by their ORDFs) is commonly solved by the Lagrangian multiplier method [191]. Again, the Lagrangian multiplier method will only find solutions which belong to the convex hull, but we are interested in all optimal solutions, which is the main reason for introducing the pruning scheme.

We need the total ORDF to be able to optimally solve the minimum distortion problem formulated in Eq. (8.28). Hence the problem is to create the total ORDF, using the M ORDFs of the boundaries. In [207, 208] a dynamic programming based approach is presented for the case where the ORDFs are defined on the set of positive integers. We introduce a different approach which does not require that the ORDFs are defined for all the positive integers, but works for ORDFs defined on any finite subset of the real line. The total ORDF can be found by applying a slightly modified version of the pruning scheme to the M different ORDFs of the boundaries. We explain this scheme with the help of the example in Fig. 8.10. In the rounded boxes on top are the points of the ORDFs of three different boundaries. The notation used in this figure is of the following form : "rate/distortion". The goal is to merge these three ORDFs to find the total ORDF, which is displayed on the right. This is achieved by

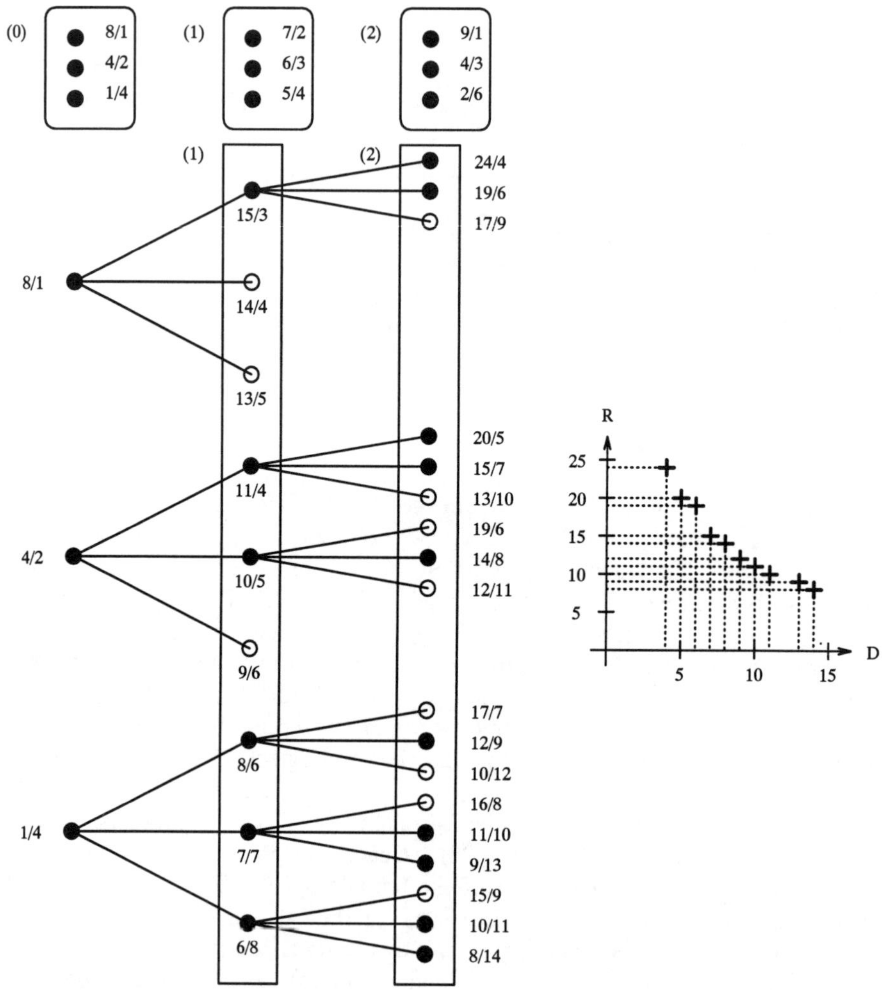

Figure 8.10 Pruned decision tree for the optimal encoding of three boundaries. The nodes are labeled as follows: "rate/distortion".

creating the total ORDF iteratively. First, we generate the combined ORDF of the first and the second boundary. This is achieved by creating all possible rate distortion points, which are inside box number one. Then we apply the same pruning rule we established before, i.e., if there are two nodes j and i such that $D(j) \geq D(i)$ and $R(j) \geq R(i)$, then node j cannot belong to the optimal solution. The pruned nodes are indicated with an empty circle and the black nodes represent the new combined ORDF. We then iteratively apply this merging and pruning of two ORDFs to create a combined ORDF, until there is only one ORDF left, which represents the total ORDF and it can directly be used to find the optimal solution to the multiple boundary encoding scheme.

8.8 VERTEX ENCODING SCHEME

So far we have not assumed any specific scheme for encoding the vertices of the polygon. In this section we present a linear and a logarithmic vertex encoding scheme which can be considered a combination of an 8-connect chain code and a run-length encoding scheme.

8.8.1 Basic linear scheme

The chain code and the run-length encoding can be combined by representing the increment between two vertices by an angle α and a run β, which form the symbol (α,β). Therefore for a run of 1, the 8 closest neighbors of a given point P are:

$$\begin{array}{ccc} (3,1) & (2,1) & (1,1) \\ (4,1) & P & (0,1) \\ (-3,1) & (-2,1) & (-1,1). \end{array} \tag{8.36}$$

As an example, $(3,4)$ represents a straight line of 4 increments in the $3 * \pi/4$ direction. Each of the possible symbols (α,β) gets a probability assigned and the resulting stream of increments $I = [(\alpha_1, \beta_1), \ldots, (\alpha_{N-1}, \beta_{N-1})]$ can be encoded by an arithmetic or a Huffman code. We use the following code word assignment. For a given symbol (α,β), the first 3 bits indicate one of the 8 possible values for α followed by $(\beta - 1)$ zeros and a final "1" to encode the number of runs. Clearly the number of bits used for this uniquely decodable code is equal to $(3 + \beta)$ and the longer the run, the more efficient this code is. Since the code word length increases linearly with the run-length, we call this the linear vertex encoding scheme. Note that this code implies that the lines between the vertices are restricted to intersect the horizontal axis in an angle which is an integer multiple of $\pi/4$.

8.8.2 Generalized linear scheme

A generalization of the linear code is based on the observation that the above scheme is optimal in the case where the probability mass function of (α, β) is separable and α is uniformly distributed over all 8 α's whereas β is geometrically distributed ($P(\beta = j) = \frac{1-\gamma}{\gamma} * \gamma^j, j \geq 1$) with parameter $\gamma = 0.5$. The assumptions that the distribution is separable, α is uniformly distributed and β is geometrically distributed are reasonable but there might be better choices for γ than 0.5. When an arithmetic coder is used, the resulting bit rate is the entropy based on the probability model the encoder employs (we neglect the renormalization bits used in practical implementations). Therefore a probability model which leads to a smaller entropy than the above can be used, even though this leads to fractional bit assignments per symbol. For example, only 6 out of the 8 α's need to be considered since the next α cannot be equal to the current one (if so, this would be coded by an additional run), nor can it be equal to the direct opposite one (if so, one less run would have been coded). Hence there are only 6 possible α's, and instead of using 3 bits to encode them only $\log_2(1/6) \approx 2.6$ bits are needed.

The question therefore is, which γ leads to the smallest bit rate for the encoding of a particular polygon. It can be shown that the maximum likelihood estimate γ_{ML} also leads to the minimum entropy, and hence to the smallest bit rate. Since we assume that the runs β_i for the encoding of vertices p_i are independent of each other and have the same geometric probability mass function, the likelihood function can be written as follows,

$$P(\beta_1, \ldots, \beta_{N_P-1}) = \prod_{i=1}^{N_P-1} \frac{1-\gamma}{\gamma} * \gamma^{\beta_i}, \tag{8.37}$$

which leads to the following maximum likelihood estimate of γ,

$$\gamma_{ML} = 1 - \frac{N_P - 1}{\sum_{i=1}^{N_P-1} \beta_i}. \tag{8.38}$$

So far we have considered the case where γ_{ML} has been given (in other words, the code word assignment has been given) and then the optimal polygon approximation is found. The question arises on how to jointly select γ_{ML} and the polygon approximation optimally. In fact γ_{ML} has to be quantized since it needs to be sent for every boundary and in the current scheme an 8 bit uniform quantizer with a range from zero to one is used for that purpose. Therefore 256 different γ's exist and one solution is to run the optimal polygon approximation 256 times and pick the quantized γ which leads to the smallest rate.

We propose a much faster, but suboptimal iterative procedure to estimate γ_{ML}. This procedure can be applied to all schemes presented, which do not employ an iteration whose convergence is based on the global optimality of the solution. In other words, this scheme can be applied to the minimum rate case for distortion measures of class one and to the pruning scheme introduced for class two distortion measures in section 8.4.2.

The iteration works as follows: first an initial quantized $\gamma^Q_{ML,i}$ is used, where i indicates the iteration number and Q the fact that this is a quantized value, and the $(i+1)$-st optimal polygon approximation is found. Then this approximation is used to estimate $\gamma_{ML,i+1}$ based on the distribution of the runs and the quantized $\gamma^Q_{ML,i+1}$ is derived from $\gamma_{ML,i+1}$. These three steps are repeated until the minimum rate for the polygon of iteration $(i+1)$, $R^*_{i+1}(\gamma^Q_{ML,i})$, does not decrease any further which usually happens after two to three iterations. Since the minimum rate of the polygonal approximation is bounded from below by 0, we can prove that this iteration converges to a local minimum by showing that $R^*_{i+1}(\gamma^Q_{ML,i})$ is a non-increasing function of $\gamma^Q_{ML,i}$.

Proof: Clearly $R^*_{i+1}(\gamma^Q_{ML,i}) \geq R^*_{i+1}(\gamma_{ML,i+1})$ since $\gamma_{ML,i+1}$ is estimated using the runs of the optimal polygon of iteration $(i+1)$. Since the likelihood function used to find $\gamma_{ML,i+1}$ is concave, $\gamma^Q_{ML,i+1}$ can be found by evaluating the likelihood function for the two reconstruction levels which are the closest to $\gamma_{ML,i+1}$ and setting $\gamma^Q_{ML,i+1}$ equal to the one which results in the higher score. Note that this is a special case where the optimal solution to a discrete problem (finding $\gamma^Q_{ML,i+1}$) can be inferred from the solution of a continuous problem (finding $\gamma_{ML,i+1}$). Therefore $R^*_{i+1}(\gamma^Q_{ML,i}) \geq R^*_{i+1}(\gamma^Q_{ML,i+1}) \geq R^*_{i+2}(\gamma^Q_{ML,i+1})$ which proves the convergence to a local minimum. ■

Using an arithmetic coder and this iterative scheme, the efficiency of the minimum rate approach for class one distortion measures can be improved by about 15%.

8.8.3 Logarithmic scheme

The linear scheme presented in the previous two sections results in a code word length which grows linearly with the run-length. It is well known from universal coding that integers can be encoded using a code which grows only logarithmically with the size of the integer.

The idea behind this universal code is that every positive integer X has a binary representation which requires $\lceil \log_2(X) \rceil$ bits, where $\lceil \cdot \rceil$ denotes the ceiling function. Clearly this binary representation can only be decoded if the decoder knows how many bits are in the binary representation, otherwise a binary stream of integers could not be separated into individual integers. Hence the length of the binary representation needs to be encoded also, which can be achieved in the same fashion, since this length is also an integer. Clearly this is a recursive scheme which is invoked until the length to be encode is 1. In the case where there is an upper limit in the size of the integer, a recursion might not be necessary since a fixed length code word can be used to indicate the length of the binary representation. This is the approach we are following in this section.

The code we propose for the run-length encoding is displayed in Table 8.1, where Y is the length of X and $[YX]$ is the code word sent. Note that the code length increases logarithmically with the run-length and that this code will result in a smaller bit rate than the linear code for runs larger than 4 and in a larger bit rate for runs smaller than 3. In other words, if there are a lot of short runs (≤ 4) the linear code is better, whereas for objects with longer runs, the logarithmic code is more efficient. Also note that in this particular logarithmic code, the longest run possible is 15.

8.8.4 Improved orientation encoding

The formulation of the optimization problem requires that the encoding cost (edge rate) from vertex p_i to vertex p_j is only a function of these two vertices. Hence every direction of departure from vertex p_i must have the same probability, which is 1/8. Clearly it takes 3 bits to encode 8 equally probable directions. On the other hand, in natural boundaries, the arrival direction of the edge (p_k, p_i) is highly correlated with the departure direction of the edge (p_i, p_j). This implies that the arrival direction should be used to predict the departure direction.

run	Y	X	total length
1	00	-	2
2	01	0	3
3	01	1	3
4	10	00	4
5	10	01	4
6	10	10	4
7	10	11	4
8	11	000	5
9	11	001	5
10	11	010	5
11	11	011	5
12	11	100	5
13	11	101	5
14	11	110	5
15	11	111	5

Table 8.1 Code word assignment for the runs

The main problem of using such a prediction is that the DAG shortest path algorithm cannot directly be applied to find the optimal polygon, since this edge direction prediction results in a higher order dependency. This problem can still be solved but it is computationally more expensive. Therefore we propose the following approximation to this edge prediction concept, which uses the same DAG shortest path algorithm we have been employing throughout this chapter.

Since we would like to use the DAG shortest path algorithm, all the departure directions need to use the same number of bits, or in other words, they should be equally likely. So far we have been using 3 bits since this is required to encode 8 equally probable directions. We propose to use only 2 bits for 4 equally probable directions. Clearly this is in conflict with the fact that there are 8 directions. The underling assumption is that 4 out of these 8 directions will never be used by the optimal solution and hence, we do not need a code word for them, even though for the optimization, they also use 2 bits.

Our experiments have shown, that the "jack-knife" directions (see Fig. 8.11a) are very unlikely to occur in an optimal solution. Therefore, even though we assume that they cost 2 bits to encode, we do not have a code word for them.

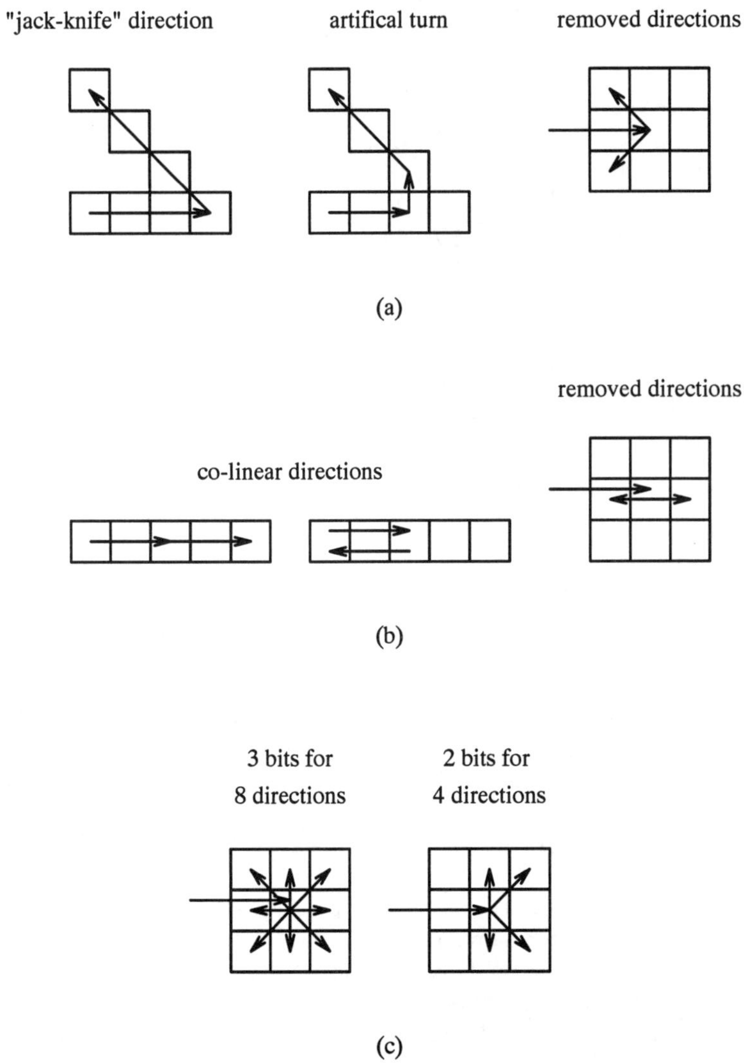

Figure 8.11 Improved orientation encoding. No code word exists for the "jack-knife" directions in Fig. (a). If a "jack-knife" direction needs to be encoded for the optimal polygon, it is approximated with an artificial turn. No optimal polygon will select the two co-linear directions displayed in Fig. (b). In Fig. (c), the original 8 directions are now encode using only 2 bits, with only four code words available.

If the optimal solution uses a "jack-knife" direction, then this direction can be obtained by adding an artificial turn, which will require a few additional bits and the distortion will change slightly (see Fig. 8.11a). In that case, the solution is not optimal, but still very close to optimal. In almost all cases however, the optimal solution will not include a "jack-knife" turn and therefore it will truly be an optimal solution.

Furthermore, an optimal solution will not append two edges which are co-linear since the compounded edge can be encoded more efficiently (see Fig. 8.11b) . Hence the two co-linear directions will never occur in an optimal polygon and hence we do not need a code word for them. This does not hold if the compounded edge is longer than the maximum run (15 in the logarithmic code). Then it might be optimal to append a co-linear edge. This case can be recognized by the encoder and decoder and treated by sending 3 bits for the next direction which allows for the next edge to be co-linear. Therefore, in most cases, the two co-linear departure directions are not needed which leaves us with 4 equally probable departure directions which in turn can be encoded by only 2 bits instead of the original 3 bits (see Fig. 8.11c).

Note that for the first vertex, all 8 departure directions are admissible since there exists no arrival direction for that vertex. Therefore, 3 bits are sent for the first vertex.

8.9 EXPERIMENTAL RESULTS

In this section we present experimental results of the proposed algorithms using object boundaries from the "Miss America" sequence. For the presented experiments, we use the linear vertex encoding scheme with $\gamma = 0.5$ and the Huffman code proposed in section 8.8. The only exception to this is the result we present at the end of this section, where we use the logarithmic run-length encoding scheme presented in section 8.8.3 and the improved orientation encoding scheme presented in 8.8.4.

We first present results for class one distortion measures, where the employed distortion measure is the minimum absolute distance. In Fig. 8.12 the original segmentation is shown, which requires 468 bits using the 8-connect chain code. In Fig. 8.13 the optimal segmentation with $D_{max} = 1$ pixel is shown, which requires a rate of 235 bits and results in a distortion of 1 pixel. The two objects in the original segmentation require 468 bits if encoded by an 8-connect chain

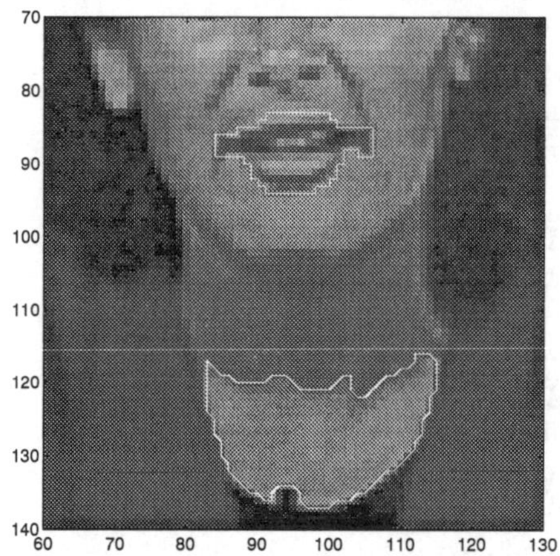

Figure 8.12 Original segmentation which requires 468 bits using the 8-connect chain code.

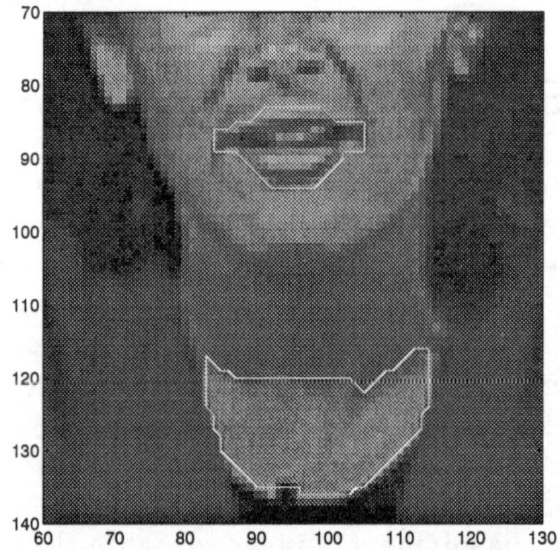

Figure 8.13 Optimal segmentation with $D_{max} = 1$ pixel, which requires a rate of 235 bits and results in a distortion of 1 pixel.

code whereas the optimal segmentation can be encoded with only 235 bits. By introducing a permissible maximum error of one pixel, we are able to reduce the total bit rate by about 50%. As expected, some of the details have been lost, i.e., the boundary has been "straightened". This smoothing of the boundary might be desired since most segmentation algorithms result in noisy boundaries. In Fig. 8.14 we show the resulting segmentation for the minimum distortion case for multiple boundaries. The maximum rate R_{max} has been set to 280 bits and the optimal solution, which uses 274 bits for a $D_{max} = 0.71$ pixels, is displayed. In Fig. 8.15 a closeup of the lower boundary in Fig. 8.14 is displayed, and the stars indicate the original boundary with the polygonal approximation drawn on top of it.

In Figs. 8.16 and 8.17 we present results for distortion measures of class two, where the employed distortion measure is the mean squared distance. To highlight the difference between the Lagrangian multiplier scheme and the pruning scheme, we set R_{max} from Eq. (8.8) to 200 bits. In Fig. 8.18 a close up of the boundary is shown and the operational rate distortion curve is displayed in Fig. 8.19. Note the possible Lagrangian solutions which are indicated by circles around the vertices of the convex hull. The Lagrangian solution which satisfies $R \leq 200$ bits results in $R = 169$ bits and $D = 0.1$ (for a $\lambda^* = 0.002$), whereas the pruning approach results in $R = 200$ bits and $D = 0.05$. The Lagrangian solution is shown in Fig. 8.16 and the solution of the pruning is shown is Fig. 8.17. It is clear that the pruning approach results in a better approximation of the original boundary.

So far we have presented results of the basic schemes, i.e., the vertices had to belong to the boundary, the linear vertex encoding scheme was used, the 8 directions were encoded using 3 bits and no secondary distortion measure was applied. In the last experiment, we use all of the above extensions for the minimum rate case for a class one distortion measure. Again, we select the maximum absolute distance as the edge distortion measure. The maximum distortion D_{max} was selected to be one pixel and we jointly optimally encode both "Miss America" objects. For the band of admissible vertices we select DM = 1 pixel, which is the same as the maximum distortion. Our experiments have shown that this is a good tradeoff between speed and coding efficiency. In Fig. 8.20 the resulting approximation is displayed, which uses only 123 bits and results in a maximum distortion of 1 pixel. Clearly this shows the power of the proposed extensions, since this is about half the bit rate required for the basic scheme (235 bits) with the same maximum distortion. The main gain comes from the fact that the polygon is permitted to use vertices which do not belong to the original boundary. This results in longer runs which are more efficiently encoded using the logarithmic run-length code. Also only 2 bits are spent for

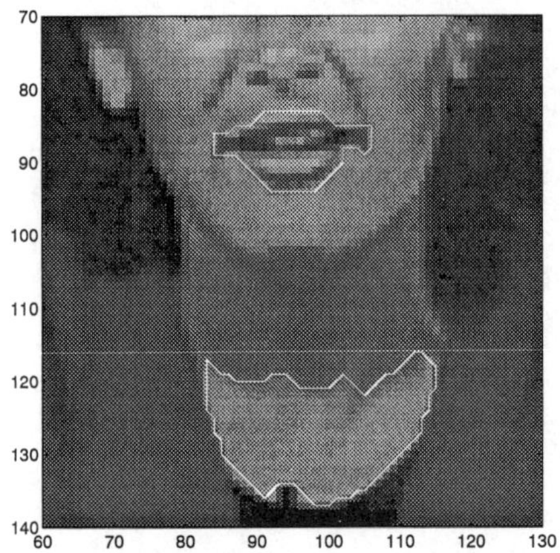

Figure 8.14 Optimal segmentation for $R_{max} = 280$ bits which results in a distortion of 0.71 pixels and a bit rate of 274 bits.

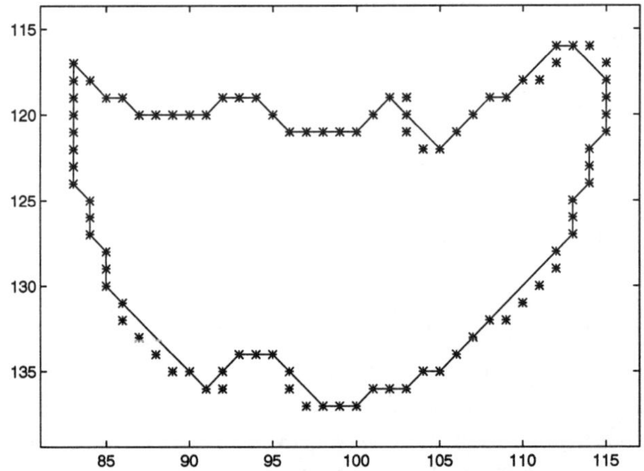

Figure 8.15 Closeup of the lower boundary; the stars indicate the original boundary and the line represents the polygonal approximation. The upper left corner has been selected as the first vertex.

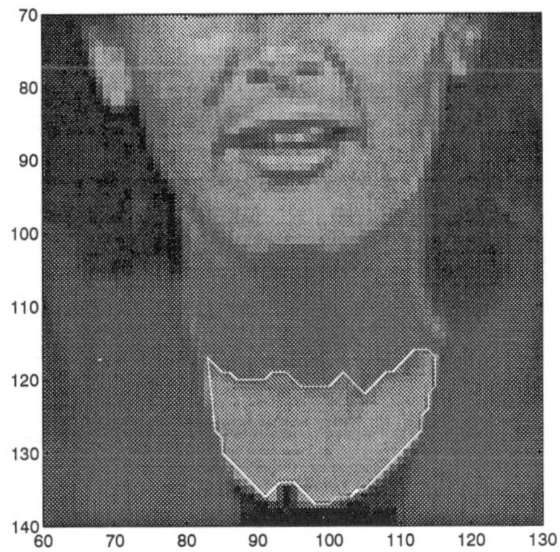

Figure 8.16 The result of the Lagrangian multiplier approach, where the target bit rate is $R_{max} = 200$ and the distortion measure is the mean squared error. The resulting bit rate is $R = 169$ bits and the resulting distortion is D=0.1.

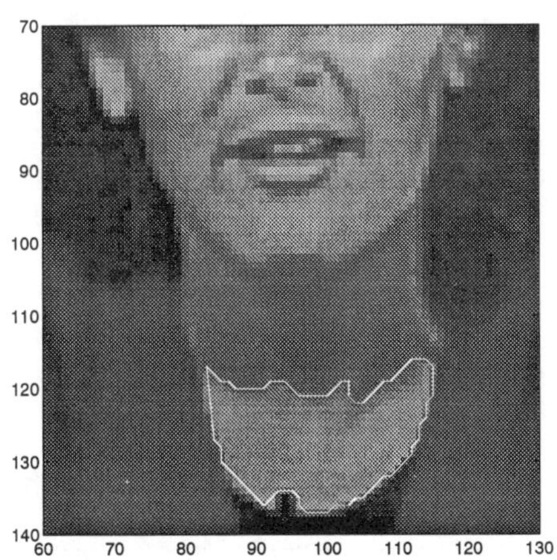

Figure 8.17 The result of the pruning approach, where the target bit rate is $R_{max} = 200$ and the distortion measure is the mean squared error. The resulting bit rate is $R = 200$ bits and the resulting distortion is D=0.05.

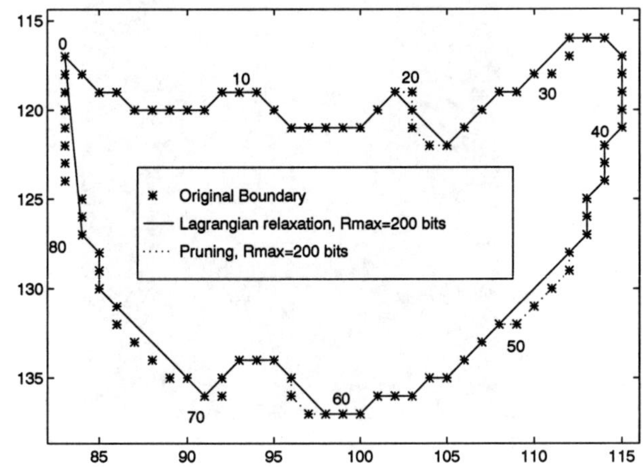

Figure 8.18 Comparison between the Lagrangian approach and the pruning approach.

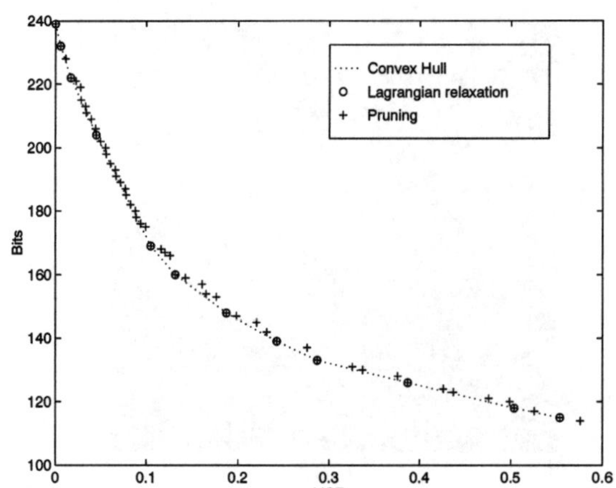

Figure 8.19 Operational rate distortion function and its convex hull.

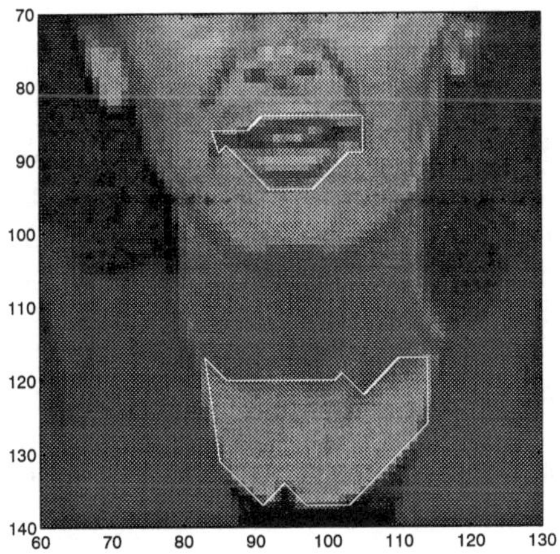

Figure 8.20 Optimal segmentation with $D_{max} = 1$ pixel, which requires a rate of 123 bits and results in a distortion of 1 pixel.

each direction using the improved orientation encoding scheme. Even though the distortion of the polygon approximation of the mouth in Fig. 8.20 is the same as in Fig. 8.13, most human observers prefer the approximation in Fig. 8.13. The reason for this is that in Fig. 8.13, the vertices have to belong to the original boundary and hence the approximation has to stay closer to the original boundary. Note that this shows that distortion measures do not always coincide with human perception, which is a well known fact.

8.10 SUMMARY

In this chapter, we presented fast and efficient methods for the lossy encoding of object boundaries which are given as 8-connect chain codes. The boundary is approximated by a polygon and we considered the problem of finding the polygon which leads to the smallest distortion for a given number of bits. The dual problem of finding the polygon which leads to the smallest bit rate for a given distortion was also addressed. We considered two different classes of distortion measures, where the first class is based on the maximum operator and the second class is based on the summation operator. For the first class, we derived a scheme which is based on a shortest path algorithm for a weighted directed acyclic graph. For the second class we proposed a Lagrangian multiplier based approach, which employs the shortest path algorithm iteratively. Lagrangian schemes can only find solutions which belong to the convex hull of the operational rate distortion function. Therefore we also proposed a tree pruning algorithm which can find all optimal solutions. We proposed a modification of the basic algorithm which allows for a secondary distortion measure. Such a measure is used to select the best of all optimal polygons, when there are more than one optimal polygon. We introduced an extension of the basic schemes to also permit polygons which use vertices which do not belong to the boundary. We extended all proposed schemes to the jointly optimal encoding of multiple boundaries. We finally introduced vertex encoding schemes which are a combination of an 8-connect chain code and a run-length scheme. Experimental results of the proposed schemes were presented using objects from the "Miss America" sequence.

REFERENCES

[1] Digital video. *IEEE Spectrum*, pages 24–30, March 1992.

[2] Practical desktop video. *Byte*, April 1992.

[3] Special report: Interactive multimedia. *IEEE Spectrum*, pages 22–39, March 1993.

[4] J. K. Aggarwal and N. Nandhakumar. On the computation of motion from sequences of images. *Proceedings of the IEEE*, 76:917–935, August 1988.

[5] N. Ahmed, T. Natarajan, and K. R. Rao. Discrete cosine transform. *IEEE Transactions on Computers*, 23:88–93, January 1974.

[6] K. Aizawa, C. S. Choi, H. Harashima, and T. S. Huang. Human facial motion analysis and synthesis with application to model-based coding. In M. I. Sezan and R. L. Lagendijk, editors, *Motion analysis and images sequence processing*. Kluwer, Norwell, MA, 1993.

[7] K. Aizawa, H. Harashima, and T. Saito. Model-based analysis-synthesis image coding (MBASIC) system for person's face. *Signal Processing: Image Communication*, 1, October 1989.

[8] K. Aizawa and T. S. Huang. Model-based image coding: advanced video coding techniques for very low bit-rate applications. *Proceedings of the IEEE*, 83(2):259–271, February 1995.

[9] C. Arnest, M. Mandel, P. Elstrom, R. Grover, E. Lesly, P. Coy, and M. Lewyn. Telecom's new age. *Business Week*, pages 63–87, April 1996. Special Report.

[10] B. Astle. *A proposal for MPEG video report, Rev. 1.* ISO-IEC/JTC1/SC2/WG11, August 1991.

[11] M. R. Banham, J. C. Brailean, C. L. Chan, and A. K. Katsaggelos. Low bit rate video coding using robust motion vector regeneration in the decoder. *IEEE Transactions on Image Processing*, 3(5):652–665, September 1994.

[12] R. E. Bellman. *Dynamic programming.* Princeton University Press, Princeton, 1957.

[13] T. Berger. *Rate distortion theory: A mathematical basis for data compression.* Prentice-Hall, 1971.

[14] D. P. Bertsekas. *Dynamic programming: Deterministic and stochastic models.* Prentice-Hall, 1987.

[15] V. Bhaskaran and K. Konstantinides. *Image and video compression standards: Algorithms and architectures.* Kluwer academic publishers, 1995.

[16] M. Bierling. Displacement estimation by hierarchical blockmatching. In *Proceedings of the Conference on Visual Communications and Image Processing*, volume 1001, pages 942–951. SPIE, 1988.

[17] D. W. Boston. Synthetic facial communication. *British Journal of Audiology*, 7:95–101, 1973.

[18] J. C. Brailean and A. K. Katsaggelos. A recursive nonstationary MAP displacement vector field estimation algorithm. *IEEE Transactions on Image Processing*, 4(4):416–429, April 1995.

[19] J. C. Brailean and A. K. Katsaggelos. Simultaneous recursive motion estimation and restoration of noisy and blurred image sequences. *IEEE Transactions on Image Processing*, 4(9):1236–1251, September 1995.

[20] J. C. Brailean, R. P. Kleihorst, S. N. Efstratiadis, A. K. Katsaggelos, and R. L. Lagendijk. Noise reduction filters for dynamic image sequences: A review. *Proceedings of the IEEE*, 83(9):1272–1292, September 1995.

[21] L. Breiman, J. H. Friedman, R. A. Olshen, and C. J. Stone. *Classification and regression trees.* The Wadsworth statistics/probability series. Wadsworth, Belmont, CA, 1984.

[22] N. M. Brooke. Computer graphics synthesis of talking faces. In G. Bailly, C. Benoit, and T.R. Sawallis, editors, *Talking machines: theories, models, and design.* Elsevier Science Publisher, 1992.

[23] N. M. Brooke and Q. Summerfield. Analysis, synthesis, and perception of visible articulatory movements. *Journal of Phonetics*, 11:63–76, 1983.

[24] H. Bruesewitz. Motion compensation with triangles. In *Proceedings of the 3rd International Workshop on 64 kbit/s Coding of Moving Video*, Rotterdam, Netherlands, September 1990.

[25] P. J. Burt and E. H. Adelson. The laplacian pyramid as a compact image code. *IEEE Transactions on Communications*, 31(4):532–540, April 1983.

[26] K. Challapali, X. Lebegue, J. S. Lim, W. H. Paik, R. Saint Girons, E. Petajan, V. Sathe, P. A. Snopko, and J. Zdepski. The grand alliance system for US HDTV. *Proceedings of the IEEE*, 83(2):158–174, February 1995.

[27] C. L. Chan and A. K. Katsaggelos. Iterative maximum likelihood displacement field estimation in quantum-limited image sequences. *IEEE Transactions on Image Processing*, 4(6):743–751, June 1995.

[28] C. S. Choi, K. Aizawa, H. Harashima, and T. Takebe. Analysis and synthesis of facial image sequences in model-based image coding. *IEEE Transactions on Circuits and Systems*, 4:257–275, September 1994.

[29] P. A. Chou, T. Lookabaugh, and R. M. Gray. Optimal pruning with applications to tree-structured source coding and modeling. *IEEE Transactions on Information Theory*, 35:299–315, March 1989.

[30] R. J. Clark. *Transform coding of images*. Academic press, 1985.

[31] M. M. Cohen and D. W. Massaro. Synthesis of visible speech. *Behavioral Research Methods and Instrumentation*, pages 260–263, 1990.

[32] M. M. Cohen and D. W. Massaro. Modeling coarticulation in synthetic visual speech. In N.M. Thalmann and D.Thalmann, editors, *Models and techniques in computer animation*. Springer-Verlag, Tokyo, 1993.

[33] Y. Cohen, M. S. Landy, and M. M. Pavel. Hierarchical coding of binary images. *IEEE Transactions on Pattern Analysis and Machine Intelligence*, PAMI-7:284–298, 1985.

[34] T. Cormen, C. Leiserson, and R. Rivest. *Introduction to algorithms*. McGraw-Hill Book Company, 1991.

[35] L. D. Davisson. Rate-distortion theory and applications. *Proceedings of the IEEE*, 60:800–808, July 1972.

[36] N. Diehl. Object-oriented motion estimation and segmentation in image sequences. *Signal Processing: Image Communication*, 3:23–56, 1991.

[37] N. Diehl. Model-based image sequence coding. In M. I. Sezan and R. L. Lagendijk, editors, *Motion analysis and images sequence processing*. Kluwer, Norwell, MA, 1993.

[38] F. Dufaux. *Multigrid block matching motion estimation for generic video coding.* PhD thesis, Swiss Federal Institute of Technology in Lausanne, 1994.

[39] T. Ebrahimi, F. Dufaux, I. Moccagatta, T. G. Campbell, and M. Kunt. A digital video codec for medium bitrate transmission. In *Proceedings of the Conference on Visual Communications and Image Processing*, volume 1605, pages 2–15. SPIE, 1991.

[40] S. N. Efstratiadis and A. K. Katsaggelos. Nonstationary AR modeling and constrained recursive estimation of the displacement field. *IEEE Transactions on Circuits and Systems for Video Technology*, 2(4):334–346, December 1992.

[41] S. N. Efstratiadis and A. K. Katsaggelos. An adaptive regularized recursive displacement estimation algorithm. *IEEE Transactions on Image Processing*, 2(3):341–352, July 1993.

[42] P. Ekman and W. V. Friesen. *Facial Action Coding System.* Consulting Psychologists Press, Palo Alto, CA., 1977.

[43] N. P. Erber. Real-time synthesis of optical lip shapes from vowel sounds. *Journal of the Acoustical Society of America*, 66:1542–1544, November 1979.

[44] N. P. Erber and C. L. DeFilippo. Voice-mouth synthesis of /pa,ba,ma/. *Journal of the Acoustical Society of America*, pages 1015–1019, 1978.

[45] H. Everett. Generalized Lagrange multiplier method for solving problems of optimum allocation of resources. *Operations Research*, 11:399–417, 1963.

[46] Expert's group on very low bit rate video telephony, ITU-Telecommunication standardization sector. *Draft Recommendation H.263.*

[47] Expert's Group on Very Low Bitrate Visual Telephony. *Video Codec Test Model, TMN4 Rev1.* ITU Telecommunication Standardization Sector, October 1994.

[48] M. L. Fisher. The Lagrangian relaxation method for solving integer programming problems. *Management Science*, 27(1):1–18, January 1981.

[49] M. L. Fisher. An applications oriented guide to Lagrangian relaxation. *Interfaces*, 15(2):10–21, March-April 1985.

[50] R. Fletcher. *Practical methods of optimization.* Wiley, Chichester; New York, second edition, 1987.

[51] R. Forchheimer and O. Fahlander. Low bit-rate coding through animation. In *Proceedings of the Picture Coding Symposium*, Davis, CA, 1983.

[52] R. Forchheimer and T. Kronander. Image coding-from waveforms to animation. *Proceedings of the International Conference on Acoustics, Speech and Signal Processing*, 37(12):2008–2023, December 1989.

[53] G. D. Forney. The Viterbi algorithm. *Proceedings of the IEEE*, 61(3):268–278, March 1973.

[54] B. Fox. Discrete optimization via marginal analysis. *Management Science*, 13(3):210–216, November 1966.

[55] H. Freeman. On the encoding of arbitrary geometric configurations. *IRE Trans. Electron. Comput.*, EC-10:260–268, June 1961.

[56] N. P. Galatsanos and A. K. Katsaggelos. Methods for choosing the regularization parameter and estimating the noise variance in image restoration and their relation. *IEEE Transactions on Image Processing*, 1:322–336, July 1992.

[57] R. G. Gallager. *Information theory and reliable communication.* Wiley, 1968.

[58] A. M. Geoffrion. Lagrangean relaxation for integer programming. *Mathematical programming study 2*, 2:82–114, 1974.

[59] C. F. Gerald and P. O. Wheatley. *Applied numerical analysis.* Addison Wesley, fourth edition, 1990.

[60] P. Gerken. Object-based analysis-synthesis coding of image sequences at very low bit rates. *IEEE Transactions on Circuits and Systems*, 4(3):228–235, June 1994.

[61] A. Gersho and R. M. Gray. *Vector quantization and signal compression.* Kluwer academic publishers, 1992.

[62] H. Gharavi. Subband coding of video signals. In J. W. Woods, editor, *Subband image coding.* Kluwer academic publishers, Boston, MA, 1991.

[63] H. Gharavi and M. Mills. Block matching motion estimation algorithms - new results. *IEEE Transactions on Circuits and Systems*, pages 649–651, May 1989.

[64] H. Gharavi and A. Tabatabai. Sub-band coding of monochrome and color images. *IEEE Transations on Circuits and Systems*, 35(2):207–214, February 1988.

[65] B. Girod. The efficiency of motion-compensating prediction for hybrid coding of video sequences. *IEEE Journal on selected Areas in Communications,*, SAC-5(7):1140–1154, August 1987.

[66] B. Girod. Motion-compensating prediction with fractional-pel accuracy. *IEEE Transactions on Communications*, 41(4):604–612, April 1993.

[67] B. Girod. Rate-constrained motion estimation. In *Proceedings of the Conference on Visual Communications and Image Processing*, volume 2308, pages 1026–1034. SPIE, 1994.

[68] R. C. Gonzalez and R. C. Woods. *Digital image processing.* Addison-Wesley, 1992.

[69] R. M. Gray and H. Abut. Full search and tree searched vector quantization of speech waveforms. In *Proceedings of the International Conference on Acoustics, Speech and Signal Processing*, pages 593–596, May 1982.

[70] S. S. Haykin. *Digital communications.* Wiley, 1988.

[71] D. R. Hill, A. Pearce, and B. Wyvill. Animating speech: an automated approach using speech synthesised by rules. *The Visual Computer*, 3(5):277–288, 1988.

[72] F. S. Hill. *Computer graphics.* Macmillan Publishing Company, 1990.

[73] M. Hötter. Object-oriented analysis-synthesis coding based on moving two-dimensional objects. *Signal Processing: Image Communication*, 2(4):409–428, December 1990.

[74] M. Hötter. Optimization of an object oriented analysis-synthesis coder based on the model of flexible 2D objects. In *Proceedings of the Picture Coding Symposium*, page 10.4, Tokyo, September 1991.

[75] M. Hötter. Optimization and efficiency of an object-oriented analysis-synthesis coder. *IEEE Transactions on Circuits and Systems*, 4(2):181–194, April 1993.

[76] M. Hötter and R. Thoma. Image segmentation based on object oriented mapping parameter estimation. *Signal Processing*, 15:315–334, 1988.

[77] H. S. Hou. A fast recursive algorithm for computing the the discrete cosine transform. *Proceedings of the International Conference on Acoustics, Speech and Signal Processing*, ASSP-35(10):1455–1461, October 1987.

[78] T. R. Hsing, C.-T. Chen, and J. A. Bellisio. Video communications and services in the copper loop. *IEEE Communications Magazine*, pages 63–68, January 1993.

[79] J. J. Y. Huang and P. M. Schultheiss. Block quantization of correlated gaussian random variables. *IEEE Transactions on Communications Systems*, 11(3):289–296, September 1963.

[80] T. S. Huang and A. N. Netravali. Motion and structure from feature correspondences: A review. *Proceedings of the IEEE*, 82(2):252–268, February 1994.

[81] D. A. Huffman. A method for the construction of minimum redundancy codes. *Proc. IRE*, 40:1098–1101, 1962.

[82] ISO/IEC CD 11172. *Coding of moving pictures and associated audio for digital storage media at up to 1.5 Mbits/sec*, December 1991.

[83] ISO/IEC CD 13818-2. *Generic coding of moving pictures and associated audio*, November 1993.

[84] ITU-T Recommendations H.261. *Video codec for audivisual services at p × 64 kbits,*.

[85] A. K. Jain. Image data compression: A review. *Proceedings of the IEEE*, 69:349–389, March 1981.

[86] A. K. Jain. *Fundamentals of digital image processing.* Prentice-Hall, 1989.

[87] J. R. Jain and A. K. Jain. Displacement measurement and its application to interframe image coding. *IEEE Transactions on Communications*, 29:1799–1808, 1981.

[88] P. Kalra, A. Mangili, N. Magnenat-Thalmann, and D. Thalmann. Smile: A multilayered facial animation system. In *Proceedings IFIP Conf. on Modeling in Computer Graphics*, pages 189–198. Springer, Tokyo, 1991.

[89] T. Kaneko and M. Okudaira. Encoding of arbitary curves based on the chain code representation. *IEEE Transactions on Communications*, COM-33(7):697–707, July 1985.

[90] G. Karlsson and M. Vetterli. Three-dimensional subband coding of video. In *Proceedings of the International Conference on Acoustics, Speech and Signal Processing*, pages 1100–1103, New York, NY, April 1988.

[91] S.-Z. Kiang, R. L. Baker, G. J. Sullivan, and C.-Y. Chiu. Recursive optimal pruning with applications to tree structured vector quantizers. *IEEE Transactions on Image Processing*, 1(2):162–169, April 1992.

[92] R. Koch. Dynamic 3-D scene analysis through synthesis feedback control. *IEEE Transactions on Pattern Analysis and Machine Intelligence*, 15(7):556–568, July 1993.

[93] J. Koplowitz. On the performance of chain codes for quantization of line drawings. *IEEE Transactions on Pattern Analysis and Machine Intelligence*, PAMI-3(2):180–185, March 1981.

[94] K. Krechmer. V.34 modems: off to a fast start? *Data communications*, pages 63–66, January 1994.

[95] C. Labit and H. Nicolas. Compact motion representation based on global features for semantic image sequence coding. In *Proceedings of the Conference on Visual Communications and Image Processing*, volume 1605, pages 697–708, Boston, USA, November 1991. SPIE.

[96] J. Lee. Optimal quadtree for variable block size motion estimation. In *Proceedings of the International Conference on Image Processing*, volume 3, pages 480–483, October 1995.

[97] J. Lee and B. Dickinson. Joint optimization of frame type selection and bit allocation for MPEG video encoders. In *Proceedings of the International Conference on Image Processing*, pages 962–966, November 1994.

[98] D. LeGall. MPEG : a video compression standard for multimedia applications. *Communications of the ACM*, 34:46–58, April 1991.

[99] D. LeGall. The MPEG video compression algorithm. *Signal Processing: Image Communication*, 4(2):129–140, April 1992.

[100] J. P. Lewis and F. I. Parke. Automated lip-synch and speech synthesis for character animation. In *Proceedings Human Factors in Computing Systems and Graphics Interface*, pages 143–147, 1987.

[101] H. Li, A. Lundmark, and R. Forchheimer. Image sequence coding at very low bit rates: A review. *IEEE Transactions on Image Processing*, 3(5):589–609, September 1994.

[102] W. Li and Y.-Q. Zhang. Vector-based signal processing and quantization for image and video compression. *Proceedings of the IEEE*, 83(2):317–335, February 1995.

[103] J. S. Lim. *Two-dimensional signal and image processing.* Prentice hall, 1990.

[104] D. W. Lin, C. T. Chen, and T. R. Hsing. Video on phone lines: Technology and applications. *Proceedings of the IEEE*, 83(2):175–193, February 1995.

[105] D. W. Lin, M.-H. Wang, and J.-J. Chen. Optimal delayed-coding of video sequences subject to a buffer size constraint. In *Proceedings of the Conference on Visual Communications and Image Processing*, volume 2094, pages 223–234. SPIE, 1993.

[106] Y. Linde, A. Buzo, and R. M. Gray. An algorithm for vector quantizer design. *IEEE Transactions on Communications*, 28:84–89, January 1980.

[107] M. L. Liou. Overview of the $p \times 64$ kbits/sec video coding standard. *Communications of the ACM*, 34:59–63, 1991.

[108] A. Lippman. HDTV sparks a digital revolution. *Byte*, January 1990.

[109] S. P. Lloyd. Least squares quantization in PCM. *IEEE Transactions on Information Theory*, 28:129–137, March 1982.

[110] D. G. Luenberger. *Optimization by vector space methods.* Wiley, New York, 1969.

[111] N. Magnenat-Thalmann, E. Primeau, and D. Thalmann. Abstract muscle action procedures for human face animation. *The Visual Computer*, 3:290–297, 1988.

[112] H. S. Malvar. *Signal processing with lapped transforms.* Artech House, MA, 1992.

[113] H. S. Malvar and D. H. Staelin. The LOT: transform coding without blocking effects. *IEEE Transactions on acoustics, speech and signal processing*, 37(4):553–559, April 1989.

[114] B. B. Mandelbrot. *The fractal geometry of nature.* W. H. Freeman and Company, San Francisco, 1983.

[115] S. Martucci. Symmetric convolution and the discrete sine and cosine transform. *IEEE Transactions of Signal Processing*, 42:1038–1051, May 1994.

[116] D. W. Massaro and M. M. Cohen. Evaluation and integration of visual and auditory information in speech perception. *Journal of Experimental Psychology: Human Perception and Performance*, 9(5):753–771, 1983.

[117] D. W. Massaro and M. M. Cohen. Perception of synthesized audible and visible speech. *Psychological Science*, 1(1):55–63, January 1990.

[118] J. Max. Quantizing for minimum distortion. *IRE Transations on Information Theory*, 6:7–12, March 1960.

[119] B. Moghaddam, K. J. Hintz, and C. V. Steward. Space-filling curves for image compression. In *Automatic Object Recognition*, volume 1471, pages 414–421. SPIE, 1991.

[120] A. A. Montgomery. Development of a model for generating synthetic animated lip shapes. *Journal of the Acoustical Society of America*, page S58, 1980. abstract.

[121] S. Morishima, K. Aizawa, and H. Harashima. An intelligent facial image coding scheme driven by speech and phoneme. In *Proceedings of the International Conference on Acoustics, Speech and Signal Processing*, 1989.

[122] S. Morishima and H. Harashima. Speech-to-image media conversion based on VQ and neural network. In *Proceedings of the International Conference on Acoustics, Speech and Signal Processing*, pages 2865–2868, 1991.

[123] F. Moscheni, F. Dufaux, and H. Nicolas. Entropy criterion for optimal bit allocation between motion and prediction error information. In *Proceedings of the Conference on Visual Communications and Image Processing*, volume 2094, pages 235–242. SPIE, 1993.

[124] H. G. Musmann. A unified point of view on video compression: a model-based perspective. In *International workshop very low bit rate video compression*, Urbana, IL, May 1993.

[125] H. G. Musmann, P. Pirsch, and H. Grallert. Advances in picture coding. *Proceedings of the IEEE*, 73:523–548, April 1985.

[126] H.G. Musmann, M. Hötter, and J. Ostermann. Object-oriented analysis-synthesis coding of moving images. *Signal Processing: Image Communication*, 1(2):117–138, October 1989.

[127] M. Nahas, H. Huitric, and M. Saintourens. Animation of a B-spline figure. *The Visual Computer*, 3:272–276, 1988.

[128] Y. Nakaya and H. Harashima. An iterative motion estimation method using triangular patches for motion compensation. In *Proceedings of the Conference on Visual Communications and Image Processing*, volume 1605, pages 546–557, Boston, MA, 1991. SPIE.

[129] Y. Nakaya and H. Harashima. Motion compensation at very low rates. In *Proceedings of the Picture Coding Symposium*, page 10.2, Lausanne, Switzerland, 1993.

[130] Y. Nakaya and H. Harashima. Motion compensation based on spatial transformations. *IEEE Transactions on Circuits and Systems*, 4(3):339–356, June 1994.

[131] N. M. Nasrabadi and R. A. King. Image coding using vector quantization: A review. *IEEE Transactions on Communications*, 36(8):957–971, August 1988.

[132] G. L. Nemhauser and L. A. Wolsey. *Integer and combinatorial optimization.* Wiley, 1988.

[133] A. N. Netravali and J. O. Limb. Picture coding: A review. *Proceedings of the IEEE*, 68:366–406, March 1980.

[134] D. Neuhoff and K. Castor. A rate and distortion analysis of chain codes for line drawings. *IEEE Transactions on Information Theory*, IT-31(1):53–68, January 1985.

[135] H. Nicolas and C. Labit. Global motion identification for image sequence analysis and coding. In *Proceedings of the International Conference on Acoustics, Speech and Signal Processing*, volume 4, pages 2825–2828, Toronto, Canada, May 1991.

[136] H. Nicolas and C. Labit. Region-based motion estimation using deterministic relaxation schemes for image sequence coding. In *Proceedings of the International Conference on Acoustics, Speech and Signal Processing*, volume 3, pages 265–268, 1992.

[137] J. Nieweglowski, T. G. Campbell, and P. Haavisto. A novel video coding scheme based on temporal prediction using digital image warping. *IEEE Transactions on Consumer Electronics*, 39(3):141–150, August 1993.

[138] J.-R. Ohm. Three-dimensional subband coding with motion compensation. *IEEE Transactions on Image Processing*, 3(5):559–571, September 1994.

[139] C. O'Malley. Videoconferencing, the big picture. *Computer Shopper*, pages 160–167, April 1996.

[140] A. V. Oppenheim and R. W. Schafer. *Discrete-time signal processing.* Prentice-Hall, Inc., Englewood Cliffs, New Jersey 07632, 1989.

[141] M. T. Orchard and A. K. Katsaggelos. Backward motion compensation for interlaced HDTV. *Signal Processing: Image Communication*, 5(5-6):487–501, December 1993.

[142] A. Ortega and K. Ramchandran. Forward-adaptive quantization with optimal overhead cost for image and video coding with applications to MPEG coders. In *SPIE*, San Jose, February 1995.

[143] A. Ortega, K. Ramchandran, and M. Vetterli. Optimal trellis-based buffered compression and fast approximations. *IEEE Transactions on Image Processing*, 3(1):26–40, January 1994.

[144] J. Ostermann. Modeling of 3D moving objects for an analysis-synthesis coder. In B. Girod, editor, *Sensing and reconstruction of three-dimensional objects*, volume 1260, pages 240–249. SPIE, Bellingham, Washington, 1990.

[145] J. Ostermann. Object-based analysis-synthesis coding (OBASIC) based on the source model of moving flexible 3D objects. *IEEE Transactions on Image Processing*, 3(5):705–711, September 1994.

[146] T. Özçelik. *A Very Low Bit Rate Video Codec.* PhD thesis, Dept. EECS, Northwestern University, December 1994.

[147] T. Özçelik, J. C. Brailean, and A. K. Katsaggelos. Image and video compression algorithms based on recovery techniques using mean field annealing. *Proceedings of the IEEE*, 83(2):304–316, February 1995.

[148] F. I. Parke. *A parametric model for human faces.* PhD thesis, Univercity of Utah, 1974.

[149] F. I. Parke. A model for human faces that allows speech synchronized animation. *Computer and Graphics*, 1:3–4, 1975.

[150] F. I. Parke. Parameterized models for facial animation. *IEEE Computer graphics and animation*, pages 61–68, November 1982.

[151] F. I. Parke. Control parameterization for facial animation. In N. M. Thalmann and D. Thalmann, editors, *Computer Animation '91.* Springer-Verlag, Tokyo, 1991.

[152] A. Pearce, B. Wyvill, G. Wyvill, and D. Hill. Speech and expression: A computer solution to face animation. In *Proceedings Graphics Interface '86*, pages 136–140, 1986.

[153] C. Pelachaud, N. I. Badler, and M. Steedman. Linguistic issues in facial animation. In N. M. Thalman and D. Thalmann, editors, *Computer Animation '91*. Springer-Verlag, Tokyo, 1991.

[154] W. B. Pennebaker and J. L. Mitchell. *JPEG still image data compression.* Van Nostrand Reinhold, New York, 1993.

[155] T. S. Perry. HDTV and the new digital television. *IEEE Spectrum*, pages 34–45, April 1995.

[156] S. M. Platt and N. I. Badler. Animating facial expressions. *IEEE Computer Graphics*, 13, August 1981.

[157] R. Prasad, J. W. Vieveen, J. H. Bons, and J. C. Arnbak. Relative vector probabilities in differential chain coded line-drawings. In *Proc. IEEE Pacific Rim Conference on Communication, Computers and Signal Processing*, pages 138–142, Victoria, Canada, June 1989.

[158] W. H. Press, S. A. Teukotsky, W. T. Vetterling, and B. P. Flannery. *Numerical recipes in C: the art of scientific computing.* Cambridge university press, second edition, 1992.

[159] J. A. Provine and R. M. Rangayyan. Lossless compression of peanoscanned images. *Journal of Electronic Imaging*, 3(2):176–181, April 1994.

[160] M. Rabbani and P. Jones. *Digital image compression techniques.* SPIE Press, Bellingham, WA, 1991.

[161] B. Ramamurthi and A. Gersho. Nonlinear space-variant postprocessing of block coded images. *Proceedings of the International Conference on Acoustics, Speech and Signal Processing*, 34(5):1258–1267, October 1986.

[162] K. Ramchandran, A. Ortega, and M. Vetterli. Bit allocation for dependent quantization with applications to MPEG video coders. In *Proceedings of the International Conference on Acoustics, Speech and Signal Processing*, pages 381–384, April 1993.

[163] K. Ramchandran, A. Ortega, and M. Vetterli. Bit allocation for dependent quantization with applications to multiresolution and MPEG video coders. *IEEE Transactions on Image Processing*, 3(5):533–545, September 1994.

[164] K. Ramchandran and M. Vetterli. Best wavelet packet bases in a rate-distortion sense. *IEEE Transactions on Image Processing*, 2(2):160–175, April 1993.

[165] K. Ramchandran and M. Vetterli. Rate-distortion optimal fast thresholding with complete JPEG/MPEG decoder compatibility. *IEEE Transactions on Image Processing*, 5(3):700–704, September 1994.

[166] K. R. Rao and P. Yip. *Discrete cosine transform: algorithms, advantages, applications.* Academic Press, Boston, 1990.

[167] H. C. Reeves and J. S. Lim. Reduction of blocking effects in image coding. *Optical Engineering*, 23(1):34–37, January 1984.

[168] J. Ribas-Corbera and D. L. Neuhoff. Optimal bit allocations for lossless video coders: motion vectors vs. difference frames. In *Proceedings of the International Conference on Image Processing*, volume 2, pages 180–183, 1995.

[169] E. A. Riskin. Optimal bit allocation via the generalized BFOS algorithm. *IEEE Transactions on Information Theory*, 37(2):400–402, March 1991.

[170] J. D. Robbins and A. N. Netravali. Recursive motion compensation: A review. In T. S. Huang, editor, *Image sequence processing and dynamic scene analysis*, pages 76–103. Springer-Verlag, 1983.

[171] A. Rosenfeld and A. Kak. *Digital picture processing.* Academic Press, 1982.

[172] M. Rydfalk. Candide: A parameterized face. Technical report, Linköping University, Dep. Elec. Eng., 1987.

[173] J. Saghri and H. Freeman. Analysis of the precision of generalized chain codes for the representation of planar curves. *IEEE Transactions on Pattern Analysis and Machine Intelligence*, PAMI-3(5):533–539, September 1981.

[174] H. Samet. The quad tree and related hierarchical data structures. *ACM computing surveys*, 16(2):188–260, June 1984.

[175] M. R. Schroeder. Self-similarity and fractals in science and art. *J. Audio Eng. Soc.*, 37(10):795–808, October 1989.

[176] G. M. Schuster and A. K. Katsaggelos. A video compression scheme with optimal bit allocation among segmentation, motion and residual error. *IEEE Transactions on Image Processing*, November 1995. to appear.

[177] G. M. Schuster and A. K. Katsaggelos. An efficient boundary encoding scheme which is optimal in the rate distortion sense. In *Proceedings of the International Conference on Image Processing*, Lausanne, Switzerland, September 1996.

[178] G. M. Schuster and A. K. Katsaggelos. Fast and efficient mode and quantizer selection in the rate distortion sense for H.263. In *Proceedings of the Conference on Visual Communications and Image Processing*, pages 784–795. SPIE, March 1996.

[179] G. M. Schuster and A. K. Katsaggelos. An optimal boundary encoding scheme in the rate distortion sense. *IEEE Transactions on Image Processing*, February 1996. submitted for publication.

[180] G. M. Schuster and A. K. Katsaggelos. An optimal lossy segmentation encoding scheme. In *Proceedings of the Conference on Visual Communications and Image Processing*, pages 1050–1061. SPIE, March 1996.

[181] G. M. Schuster and A. K. Katsaggelos. An optimal quad-tree-based motion estimator. In *Proceedings of the Conference on Digital Compression Technologies & Systems for Video Communications*, October 1996.

[182] G. M. Schuster and A. K. Katsaggelos. An optimal segmentation encoding scheme in the rate-distortion sense. In *Proceedings of the International Symposium on Circuits and Systems*, volume 2, pages 640–643, Atlanta, GA, May 1996.

[183] G. M. Schuster and A. K. Katsaggelos. A very low bit rate video codec with optimal trade-off among DVF, DFD and segmentation. In *Proceedings of the European Signal Processing Conference*, Trieste, Italy, September 1996. invited paper.

[184] G. M. Schuster and A. K. Katsaggelos. A video compression scheme with optimal bit allocation between displacement vector field and displaced frame difference. In *Proceedings of the International Conference on Acoustics, Speech and Signal Processing*, May 1996.

[185] G. M. Schuster and A. K. Katsaggelos. A video compression scheme with optimal bit allocation between displacement vector field and displaced frame difference. *IEEE Journal on selected areas in communications*, September 1996. submitted for publication.

[186] V. Seferidis and M. Ghanbari. General approach to block-matching motion estimation. *Optical Engineering*, 32:1464–1474, July 1993.

[187] A. Segall. Bit allocation and encoding for vector sources. *IEEE Transactions on Information Theory*, 22(2):162–169, March 1976.

[188] C. E. Shannon. A mathematical theory of communications. *Bell system tech. journal*, 27:397–423, 623–656, 1948.

[189] C. E. Shannon. Coding theorems for a discrete source with a fidelity criterion. In *IRE Nat. Conv. Rec.*, volume 4, pages 142–163, 1959.

[190] Y. Shoham and A. Gersho. Efficient codebook allocation for an arbitrary set of vector quantizers. In *Proceedings of the International Conference on Acoustics, Speech and Signal Processing*, pages 1696–1699, Tampa, Florida, March 1985.

[191] Y. Shoham and A. Gersho. Efficient bit allocation for an arbitrary set of quantizers. *IEEE Transactions on Acoustics, Speech and Signal Processing*, 36(9):1445–1453, September 1988.

[192] E. Shusterman and M. Feder. Image compression via improved quadtree decomposition algorithms. *IEEE Transactions on Image Processing*, 3(2):207–215, March 1994.

[193] R. Smith. An approach to region based motion estimation. Master's thesis, Northwestern University, December 1995. Department of Electrical Engineering and Computer Science.

[194] G. Strang. *Linear algebra and its applications.* Harcourt Brace Jovanovich, third edition, 1988.

[195] P. Strobach. Tree-structured scene adaptive coder. *IEEE Transactions on Communications*, 38(4), April 1990.

[196] P. Strobach. Quadtree-structured recursive plane decomposition coding of images. *IEEE Transactions on Signal Processing*, 39(6):1380–1397, June 1991.

[197] G. J. Sullivan. Optimal entropy constrained scalar quantization for exponential and laplacian random variables. In *Proceedings of the International Conference on Acoustics, Speech and Signal Processing*, volume 5, pages 265–268, 1994.

[198] G. J. Sullivan and R. L. Baker. Efficient quadtree coding of images and video. In *Proceedings of the International Conference on Acoustics, Speech and Signal Processing*, pages 2661–2664, May 1991.

[199] G. J. Sullivan and R. L. Baker. Motion compensation for video compression using control grid interpolation. In *Proceedings of the International Conference on Acoustics, Speech and Signal Processing*, pages 2713–2716, May 1991.

[200] G. J. Sullivan and R. L. Baker. Efficient quadtree coding of images and video. *IEEE Transactions on Image Processing*, 3(3):327–331, May 1994.

[201] J. Sutherland and L. Litteral. Residential video services. *IEEE Communications Magazine*, pages 37–41, July 1992.

[202] M. A. Tekalp. *Digital video processing.* Prentice Hall, 1995.

[203] D. Terzopoulos and K. Waters. Analysis of facial images using and anatomical models. In *Proceedings of the International Conference on Computer Vision*, pages 727–732, 1990.

[204] D. Terzopoulos and K. Waters. Physically-based facial modeling, analysis and animation. *Journal of Visualization and Computer Animation*, 1(4):73–80, 1990.

[205] D. Terzopoulos and K. Waters. Techniques for realistic facial modeling and animation. In N. M. Thalman and D. Thalmann, editors, *Computer Animation '91.* Springer-Verlag, Tokyo, 1991.

[206] T.Özçelik and A. K. Katsaggelos. Detection and encoding of model failures in very low bit-rate video coding. In *Proceedings of the Conference on Visual Communications and Image Processing*, pages 820–831, Orlando, FL, March 1996. SPIE.

[207] A. V. Trushkin. Bit number distribution upon quantization of a multivariate random variable. *Problems of Information Transmission*, 16:76–79, 1980. translated from Russian.

[208] A. V. Trushkin. Optimal bit allocation algorithm for quantizing a random vector. *Problems of Information Transmission*, 17:156–161, 1981. translated from Russian.

[209] R. Y. Tsai and T. S. Huang. Estimating three-dimensional motion parameters of a rigid planar patch. *IEEE Transactions on Acoustics, Speech and Signal Processing*, ASSP-29(6):1147–1152, December 1981.

[210] J. Vaisey and A. Gersho. Image compression with variable block size segmentation. *IEEE transations on signal processing*, 40(8):2040–2060, August 1992.

[211] M. Vetterli. Multidimensional subband coding: Some theory and algorithms. *Signal processing*, 6(2):97–112, April 1984.

[212] M. Vetterli and J. Kovačević. *Wavelets and subband coding.* Prentice Hall, 1995.

[213] M. Vetterli and H. J. Nussbaumer. Simple FFT and DCT algorithms with reduced number of operations. *Signal Processsing*, 6(4):267–278, August 1984.

[214] M. Vetterli and K. M. Uz. Multiresolution coding techniques for digital television: A review. *Multidimensional systems and signal processing*, 3:161–187, 1992.

[215] D. R. Walker and K. R. Rao. Improved pel-recursive motion compensation. *IEEE Transactions on Communications*, COM-32:1128–1134, October 1984.

[216] G. K. Wallace. The JPEG picture compression standard. *IEEE Transactions on Consumer Electronics*, 38(1):18–34, February 1992.

[217] Y. Wang and O. Lee. Active mesh - a feature seeking and tracking image sequence representation scheme. *IEEE Transactions on Image Processing*, 3(5):610–624, September 1994.

[218] K. Waters. A muscle model for animating three-dimensional facial expression. *ACM Computer Graphics*, 21(4):17–24, July 1987.

[219] K. Waters and D. Terzopoulos. The computer synthesis of expressive faces. *Phil. Trans. R. Soc. Lond.*, 335:87–93, 1992.

[220] P. H. Westerink, J. Biemond, and D. E. Boekee. Sub-band coding of images using predictive vector quantization. In *Proceedings of the International Conference on Acoustics, Speech and Signal Processing*, pages 1378–1381, Dallas, Texas, April 1987.

[221] P. H. Westerink, J. Biemond, and D. E. Boekee. Evaluation of image subband coding schemes. In *Proceedings of the European Signal Processing Conference*, pages 1149–1152, Grenoble, France, September 1988.

[222] P. H. Westerink, J. Biemond, and D. E. Boekee. An optimal bit allocation algorithm for sub-band coding. In *Proceedings of the International Conference on Acoustics, Speech and Signal Processing*, pages 757–760, 1988.

[223] P. H. Westerink, J. Biemond, D. E. Boekee, and J. W. Woods. Subband coding of images using vector quantization. *IEEE Transactions on Communications*, 36:713–719, June 1988.

[224] T. Wiegand, M. Lightstone, T. G. Campbell, and S. K. Mitra. Efficient mode selection for block based motion compensated video coding. In *Proceedings of the International Conference on Image Processing*, volume 2, pages 559–562, October 1995.

[225] T. Wiegand, M. Lightstone, T. G. Campbell, and S. K. Mitra. A rate-constrained encoding strategy for H.263 video compression. In *Proceedings of the International Symposium on Multimedia Communications and Video Coding*, Brooklyn, NY, October 1995.

[226] T. Wiegand, M. Lightstone, D. Mukherjee, T. G. Campbell, and S. K. Mitra. Rate-distortion optimized mode selection for very low bit rate video coding and the emerging H.263 standard. *Transactions on Circuits and Systems for Video Technology*, 6(2):182–190, April 1996.

[227] I. H. Witten, R. M. Neal, and J. G. Cleary. Arithmetic coding for data compression. *Communications of the ACM*, 30:520–540, 1987.

[228] R. C. Wood. On optimum quantization. *IEEE Transactions on Information Theory*, 15(2):248–252, February 1969.

[229] J. W. Woods, editor. *Subband image coding.* Kluwer academic publishers, Boston, MA, 1991.

[230] J. W. Woods and S. D. O'Neil. Sub-band coding of images. *IEEE Transactions on Acoustics, Speech and Signal Processing*, ASSP-34:1278–1288, October 1986.

[231] Y. Yang, N. P. Galatsanos, and A. K. Katsaggelos. Regularized image reconstruction from incomplete block discrete cosine transform data. *IEEE Transactions on Circuits and Systems for Video Technology*, 3(6):421–432, December 1993.

[232] Y. Yang, N. P. Galatsanos, and A. K. Katsaggelos. Set theoretic spatially-adaptive reconstruction of block transform compressed images. *IEEE Transactions on Image Processing*, 4(7):896–908, July 1995.

[233] J. F. S. Yau. A texture mapping approach to 3-D facial image synthesis. *Computer Graphics Forum*, 7(2):129–134, 1988.

[234] X. Zhang, M. C. Cavenor, and J. F. Arnold. Adaptive quadtree coding of motion compensated image sequences for the use on the broadband ISDN. *IEEE Transactions on Circuits and Systems*, 3(3):222–229, June 1993.

INDEX

INDEX

Q

R

T

V